GeoPlanet: Earth and Planetary Sciences

More information about this series at http://www.springer.com/series/8821

Tymon Zielinski · Marcin Weslawski
Karol Kuliński
Editors

Impact of Climate Changes on Marine Environments

Springer

Editors
Tymon Zielinski
Air-Sea Interactions
Institute of Oceanology
Polish Academy of Sciences
Sopot
Poland

Karol Kuliński
Marine Chemistry and Biochemistry
Institute of Oceanology
Polish Academy of Sciences
Sopot
Poland

Marcin Weslawski
Marine Ecology
Institute of Oceanology
Polish Academy of Sciences
Sopot
Poland

The GeoPlanet: Earth and Planetary Sciences Book Series is in part a continuation of Monographic Volumes of Publications of the Institute of Geophysics, Polish Academy of Sciences, the journal published since 1962 (http://pub.igf.edu.pl/index.php).

ISSN 2190-5193 ISSN 2190-5207 (electronic)
GeoPlanet: Earth and Planetary Sciences
ISBN 978-3-319-14282-1 ISBN 978-3-319-14283-8 (eBook)
DOI 10.1007/978-3-319-14283-8

Library of Congress Control Number: 2014958026

Springer Cham Heidelberg New York Dordrecht London

Printed on acid-free paper

Springer International Publishing AG Switzerland is part of Springer Science+Business Media (www.springer.com)

Series Editors

Managing Editor

Advisory Board

Kon-Kee Liu
Institute of Hydrological
and Oceanic Sciences
National Central University Jhongli
Jhongli, Taiwan

Teresa Madeyska
Research Centre in Warsaw
Institute of Geological Sciences
Warszawa, Poland

Stanisław Massel
Institute of Oceanology
Polish Academy of Sciences
Sopot, Polska

Antonio Meloni
Instituto Nazionale di Geofisica
Rome, Italy

Evangelos Papathanassiou
Hellenic Centre for Marine Research
Anavissos, Greece

Kaja Pietsch
AGH University of Science
and Technology
Kraków, Poland

Dušan Plašienka
Prírodovedecká fakulta, UK
Univerzita Komenského
Bratislava, Slovakia

Barbara Popielawska
Space Research Centre
Polish Academy of Sciences
Warszawa, Poland

Tilman Spohn
Deutsches Zentrum für Luftund
Raumfahrt in der Helmholtz
Gemeinschaft
Institut für Planetenforschung
Berlin, Germany

Krzysztof Stasiewicz
Swedish Institute of Space Physics
Uppsala, Sweden

Roman Teisseyre
Earth's Interior Dynamics Lab
Institute of Geophysics
Polish Academy of Sciences
Warszawa, Poland

Jacek Tronczynski
Laboratory of Biogeochemistry
of Organic Contaminants
IFREMER DCN_BE
Nantes, France

Steve Wallis
School of the Built Environment
Heriot-Watt University
Riccarton, Edinburgh
Scotland, UK

Wacław M. Zuberek
Department of Applied Geology
University of Silesia
Sosnowiec, Poland

Acknowledgments

This book has been partly prepared within the framework of the activities of the Centre for Polar Studies National Leading Research Centre, 60 Bedzinska Street, 41-200 Sosnowiec, Poland.

Contents

Introduction

Tymon Zielinski, Marcin Weslawski, Karol Kuliński, Tomasz Petelski and Beata Szymczycha

Climate has a large influence on marine environments. Thus, any shifts in global or regional climate result in more or less significant changes in structure and functioning of the marine ecosystems. The recently observed high anthropogenic pressure causes that climate changes and thus also alterations in marine ecosystems are faster and more distinct than before. Moreover, most of the available predictions suggest that these changes will be observed also in the future and their amplitude may even increase (IPCC 2013).

There are a variety of ways to facilitate understanding how various drivers, both natural and anthropogenic substances and processes that alter the Earth's energy budget, contribute to climate change. In principle, observations of the climate response to a single factor could directly show the impact of that factor, or climate models could be used to study the impact of any single factor.

Radiative forcing (RF) quantifies the change in energy fluxes caused by changes in these drivers (IPCC 2013). Atmospheric aerosols are among the most crucial drivers which effect the RF in the atmosphere, through cloud adjustments, which due to aerosols, is –0.9 [−1.9 to −0.1] W m^{-2} (medium confidence). This results in a negative forcing from most aerosols and a positive contribution from black carbon absorption of solar radiation. There is high confidence that aerosols and their interactions with clouds have offset a substantial portion of global mean forcing from well-mixed greenhouse gases. They continue to contribute the largest uncertainty to the total RF estimate (IPCC 2013).

Aerosol particles affect atmospheric processes. After deposition to the sea surface, they affect processes in sea water. Their effects in the atmosphere are very

T. Zielinski (✉) · M. Weslawski · K. Kuliński · T. Petelski · B. Szymczycha
Institute of Oceanology Polish Academy of Sciences, Warsaw, Poland
e-mail: tymon@iopan.gda.pl

T. Zielinski · K. Kuliński
Centre for Polar Studies National Leading Research Centre, 60 Będzińska Street, 41-200 Sosnowiec, Poland

T. Zielinski et al. (eds.), *Impact of Climate Changes on Marine Environments*, GeoPlanet: Earth and Planetary Sciences, DOI 10.1007/978-3-319-14283-8_1

diverse and depend on the chemical and physical properties of the aerosol particles. Aerosols have a strong impact on climate both due to scattering and absorption of incoming solar radiation (direct effect) and through their effects on cloud properties and associated cloud albedo (first indirect effect) and precipitation (second indirect effect). Aerosol radiative forcing is a critical, however, variable and still quite unrecognized, component of the global climate, which results in the fact that climate models have to rely on incomplete information of the aerosol optical properties (Eck et al. 2010). The multiyear, multi-instrument observations show robust differentiation in both the magnitude and spectral dependence of the absorption—a property driving aerosol climate forcing, for desert dust, biomass burning, urban–industrial, and marine aerosols (Dubovik et al. 2002). Moreover, we observe significant variability of the absorption for the same aerosol type due to different meteorological and source characteristics as well as different emission characteristics.

Atmospheric aerosols consist of a suspension of particles whether they occur as particles or as droplets, i.e. chemicals in their liquid phase, or dissolved in a liquid in the air. Aerosols are an important constituent of the atmospheric boundary layer. Aerosol particles provide surfaces for heterogeneous chemical processes, they also act as a condensation sink for atmospheric trace gases, while hygroscopic particles serve as cloud condensation nuclei. Aerosol particle sizes vary from a few nanometres to some tens of micrometres and the "large" particles are sufficiently heavy that their atmospheric residence time is very short and hence their concentrations are negligible. Still in very strong wind conditions (hurricanes) these large sea spray particles may be important in the ocean–atmosphere transfer of heat and water vapour (Andreas et al. 2008).

As a result of various interacting processes, the most abundant aerosol particles in the atmosphere are those with a radius of a few tenths of microns, which are often referred to as accumulation mode particles. Their atmospheric lifetime is relatively long (few days to a few weeks) and it depends on their surface roughness and related deposition velocity, and their main removal mechanism is wet deposition.

Atmospheric aerosols originate from a wide variety of sources in both marine and continental environments. Aerosol content varies significantly depending upon whether the air mass is natural or modified anthropogenically, marine or continental, rural or urban (Zielinski and Zielinski 2002). Aerosols formed over land by either primary or secondary formation processes are transported over the oceans and contribute substantially to the aerosol concentrations over the oceans (Kastendeuch and Najjar 2003; Smirnov et al. 2003; Zielinski 2004). Sea spray aerosol is directly produced at the sea surface through the interactions between wind and surface waves (Hobbs 2000). Ship emissions and volcanoes also contribute to primary aerosol in the marine atmosphere.

Secondary aerosol formation from gases released from the sea surface also contribute significantly to marine atmosphere aerosol loading. Estimates of the mass concentrations show that the largest aerosol contributions on a global scale are from sea spray aerosol and desert dust (Andreae and Rosenfeld 2008; Jickells et al. 2005).

The chemical and physical properties of aerosols vary both in space and time and depend on the proximity of sources and sinks. Additionally they strongly depend on the chemical and physical transformation during their lifetime. Aerosol sources are numerous and can be of natural or anthropogenic origin. They can occur in the atmosphere as particles (primary aerosols) or can be formed in the atmosphere from their precursors in the gas phase through physical and chemical reactions (secondary aerosol formation). Physical processes, in particular the vertical transport and removal of particles by dry deposition to the surface, depend on particle size and thus their deposition velocity. Very small particles are involved in the processes of growth by condensation and coagulation. Their atmospheric transport is based on turbulence and Brownian diffusion. Very large particles are subject to gravitational forces resulting in rapid sedimentation. These processes, plus the formation by direct emission and secondary formation, chemical transformations and in-cloud processing, have decisive influence on the number concentrations of aerosol particles. These concentrations differ by 10 orders of magnitude.

Aerosol particle size is one of the fundamental quantities needed to determine the role of aerosols in climate forcing, modifying the hydrological cycle, and affecting human health and to separate natural from man-made aerosol components (Viana et al. 2014). Information on aerosol sizes is retrieved from remote-sensing instruments including satellite sensors such as Moderate Resolution Imaging Spectroradiometer (MODIS) and ground-based radiometers such as Aerosol Robotic Network (AERONET) (Kleidman et al. 2005; Christopher et al. 2006; Yu et al. 2009). The role of aerosols in climate change from the regional perspective is described in the paper by Strzalkowska et al., while Pakszys et al. write about the climate implications connected with the Arctic aerosols. Markuszewski, on the other hand, pays attention to the quantification of the role of sea salt aerosols.

Climate changes influence largely the biogeochemical cycles in the marine environment. Among the most important, climate related drivers affecting dynamics and structure of the biogeochemical processes are: temperature increase, changes in the atmospheric circulation and hydrological regimes, intensification of storm surges etc. (Emerson and Hedges 2008; IPCC 2013). Observed (and predicted in the future) increase of seawater temperature affects directly many biogeochemical processes taking place in seawater, usually by accelerating them. The good example is respiration, which, to some degree, is directly proportional to the temperature. Seawater temperature is responsible also for the solubility of gaseous substances, e.g. those crucial for marine biota: O_2 and CO_2. CO_2 being product of respiration and substrate in photosynthesis is at the same time one of the most important drivers of present climate changes. Moreover, CO_2 behaves like a weak acid in seawater and thus contributes to pH—an important factor controlling number of biogeochemical cycles and determining existence and functioning of marine organisms. As a consequence, there is a complex net of feedback mechanisms between climate changes and their effects and biogeochemical functioning of the marine ecosystems. This example indicates that there is a need for more holistic approaches in studying marine environment in the context of climate and global changes (Emerson and Hedges 2008).

Another consequence of climate changes for marine biogeochemistry are regional shifts in hydrological regimes (Dragoni and Sukhija 2008). Since water is the major medium for transportation of chemical substances, changes in precipitation (in both intensity and total volume) and thus also in river runoff and groundwater discharge modify local biogeochemical cycles. This is extremely important for coastal and shelf ecosystems being under high influence of processes occurring in their catchments, like e.g. for the Baltic Sea. Changes in freshwater supply may result in direct modification of loads of different substances, including nutrients and pollutants—both extremely important for marine biota. Indirectly they may also change local biogeochemistry by strengthening or weakening stratification in the basins, which influence number of biogeochemical processes and mechanisms, e.g. determine O_2 availability in the deeper water column and on sediments surface (BACC Author Team 2008; IPCC 2013).

Climate changes may also influence the transport pathways of contaminants. Among the hazardous substances the most important for the marine ecosystems are: SO_x and NO_x, persistent organic pollutants, heavy metals and radionuclides. Contaminants may enter marine environment through local sources, like rivers, or be deposited from the atmosphere. The latter mechanisms caused that contamination becomes a global problem. However, contaminants have usually long residence time and may accumulate in the food web. Thus, these substances, which enter the ocean locally may be also transported over long distances. Both these global transport pathways allowed the contaminants to reach e.g. pristine polar environments. Thus any changes in the atmospheric circulation or water currents may cause a threat for the vulnerable marine environments that have so far been less affected by the contamination (Bard 1999).

Global Change naturally has also complex and difficult implications for the living organisms including human health. Major biologically driven problems connected with Global Change in the marine biota are biodiversity loss, species distributions shifts, alterations in the food web, drop of the useful living resources. For humans directly it is health risks (pathogens, contamination) as well as subjective wellbeing, behavior changes and stress level increase connected with changeable meteorological conditions (Depledge and Bird 2009; EMB 2013).

While biodiversity loss per se (i.e. species extinction) is a very disputable problem in the sea (there are no more than 20 proven contemporary extinctions of marine species—see Dulvy et al. 2003), the problem of overfishing and overexploitation of living resources is serious and raise a major concern (Rosenberg 2003; Harnik et al. 2012). The global change use to be connected with the invasions and threat from alien species on land, however seriousness of this problem remains disputable (Gurevitch and Padilla 2004).

The species distribution shifts (movement of thermophilic species north and retreat of cold water forms) are well described for the zooplankton communities (Beaugrand et al. 2002), yet there is a growing amount of data on the coastal benthic species (MARLIN program in UK and examples from Arctic in Weslawski et al. 2011). ICES (2013) report on the movement of commercial fish species north (e.g. cod, herring and mackerel in NE Atlantic—Corten and Lindley 2003). Still the most concern shall be given for the backbone of marine ecosystem—the microbial life—both the primary producers and microbial loop of decomposers and grazers, that are first to react on the physical changes in the environment (Berglund et al. 2005).

References

Andreas EL, Persson POG, Hare JE (2008) A bulk turbulent air-sea flux algorithm for high-wind, spray conditions. J Phys Oceanogr 38(7):1581–1596

Andreae MO, Rosenfeld D (2008) Aerosol–cloud–precipitation interactions. Part 1. The nature and sources of cloud-active aerosols. Earth-Science Reviews, 89(1):13–41

BACC Author Team (2008) Assessment of climate change for the Baltic Sea Basin. Springer, Berlin, 473 pp

Bard SM (1999) Global transport of anthropogenic contaminants and the consequences for the Arctic marine ecosystem. Mar Pollut Bull 38(5):356–379

Beaugrand G, Reid PC, Ibanez F, Lindley JA, Edwards M (2002) Reorganization of North Atlantic marine copepod biodiversity and climate. Science 96:1692–1694

Berglund J, Samuelsson K, Kull T, Müren U, Andersson A (2005) Relative strength of resource and predation limitation of heterotrophic nanoflagellates in a low-productive sea area. J Plankton Res 27:923–935

Christopher SA, Zhang JL, Kaufman YJ, Remer LA (2006) Satellite-based assessment of top of atmosphere anthropogenic aerosol radiative forcing over cloud-free oceans. Geophys Res Lett 33(15):L15816

Corten A, Lindley JA (2003) The use of CPR data in fisheries research. Prog Oceanogr 58:285–300

Depledge MH, Bird W (2009) The Blue Gym: health and wellbeing from our coasts. Mar Pollut Bull 58(7):947–948. doi:10.1016/j.marpolbul.2009.04.019 (Epub 2009 May 23)

Dragoni W, Sukhija BS (2008) Climate change and groundwater: a short review. Geol Soc London Spec Publ 288:1–12. doi:10.1122/SP288.1

Dubovik O, Holben B, Eck TF, Smirnov A, Kaufman YJ, King MD, Tanré D, Slutsker I (2002) Variability of absorption and optical properties of key aerosol types observed in worldwide locations. J Atmos Sci 59(3):590–608. doi:10.1175/1520-0469(2002)059<0590:VOAAOP>2.0.CO;2

Dulvy NK, Sadovy Y, Reynolds JD (2003) Extinction vulnerability in marine populations. Fish Fish 4:25–64

Eck TF, Holben BN, Sinyuk A, Pinker RT, Goloub P, Chen H, Chatenet B, Li Z, Singh RP, Tripathi SN, Reid JS, Giles DM, Dubovik O, O'Neill NT, Smirnov A, Wang P, Xia X (2010) Climatological aspects of the optical properties of fine/coarse mode aerosol mixtures. J Geophys Res 115:D19205. doi:10.1029/2010JD014002

EMB-European Marine Board (2013) Linking oceans and human health: a strategic research priority for Europe. Position paper 19 of the European Marine Board, Ostend, Belgium

Emerson SR, Hedges JI (2008) Chemical oceanography and the marine carbon cycle. Cambridge University Press, Cambridge, p 453

Gurevitch J, Padilla DK (2004) Are invasive species a major cause of extinctions? TREE 19:470–474

Harnik et al (2012) Extinctions in ancient and modern seas. TREE 27:608–617

Hobbs RJ (Ed) (2000) Invasive species in a changing world. Island Press

ICES (2013) Report of the ad hoc group on the distribution and migration of Northeast Atlantic Mackerel (AGDMM). ICES CM 2013/ACOM:58, 215 pp

IPCC (2013) Climate change 2013: the physical science basis. In: Stocker TF, Qin D, Plattner G-K, Tignor M, Allen SK, Boschung J, Nauels A, Xia Y, Bex V, Midgley PM (eds) Contribution of working group I to the fifth assessment report of the intergovernmental panel on climate change. Cambridge University Press, Cambridge, 1535 pp

Jickells TD, An ZS, Andersen KK, Baker AR, Bergametti G, Brooks N, Cao JJ, Boyd PW, Duce RA, Hunter Kubilay N, laRoche J, Liss PS, Mahowald N, Prospero JM, Ridgwell AJ, Torres R (2005) Global iron connections between desert dust, ocean biogeochemistry, and climate. Science 308(5718):67–71

Kastendeuch PP, Najjar G (2003) Upper-air wind profiles investigation for tropospheric circulation study. Theor appl climatol 75(3–4):149–165

Kleidman RG, O'Neill NT, Remer LA, Kaufman YJ, Eck TF, Tanre D, Dubovik O, Holben BN (2005) Comparison of moderate resolution imaging spectroradiometer (MODIS) and aerosol robotic network (AERONET) remote-sensing retrievals of aerosol fine mode fraction over ocean. J Geophys Res Atmos 110(D22):D22205

MARLIN. http://www.marlin.ac.uk/bacs.php

Rosenberg AA (2003) Managing to the margins: the over exploitation of fisheries. Front Ecol Environ 1(2):102–106

Smirnov A, Holben BN, Dubovik O, Frouin R, Eck TF, Slutsker I (2003) Maritime component in aerosol optical models derived from Aerosol Robotic Network data. J Geophys Res: Atm (1984–2012), 108(D1), AAC-14

Viana M, Hammingh P, Colette A, Querol X, Degraeuwe B, Vlieger ID, van Aardenne J (2014) Impact of maritime transport emissions on coastal air quality in Europe. Atmos Environ 90:96e105

Węsławski JM, Kendall MA, Włodarska-Kowalczuk M, Iken K, Kędra M, Legezynska J, Sejr MK (2011) Climate change effects on Arctic fjord and coastal macrobenthic diversity observations and predictions. Mar Biodiv 41:71–85

Yu HB, Chin M, Remer LA, Kleidman RG, Bellouin N, Bian HS, Diehl T (2009) Variability of marine aerosol fine-mode fraction and estimates of anthropogenic aerosol component over cloud-free oceans from the moderate resolution imaging spectroradiometer (MODIS). J Geophys Res Atmos 114:D10206

Zielinski T (2004) Studies of aerosol physical properties in coastal areas. Aerosol Sci Technol 38(5):513–524

Zielinski T, Zielinski A (2002) Aerosol extinction and optical thickness in the atmosphere over the Baltic Sea determined with lidar. J Aerosol Sci 33(6):47–61

Impact of Breaking Waves on Sea Salt Production and Local Change of Aerosol Optical Properties

A. Strzalkowska, T. Zielinski, P. Makuch, P. Pakszys and T. Petelski

Abstract In this paper we discuss local impact of breaking waves on production of sea salt aerosols and hence on the change of aerosol size distribution and particle optical properties. Our studies were made between 17 and 27 July 2012 at the Coastal Research Station (CRS) in Lubiatowo on the Polish Baltic coast. During the studies aerosol optical depth was measured using Microtops II sun photometers and AERONET and MODIS data were used to support the further analyses. We show that with the local wave breaking phenomenon the aerosol optical depth may increase by a magnitude of even one order and that the ensemble of aerosol particles may shift from the dominating fine mode to coarse mode (sea salt). Such shift may have a strong local impact on the radiative forcing and hence on a local climate.

Keywords Aerosol · Baltic · Microtops · AERONET

1 Introduction

Atmospheric aerosols originate from a wide variety of sources in both marine and continental environments. Aerosol content varies significantly depending upon whether the air mass is natural or modified anthropogenically, marine or continental, rural or urban (Zielinski and Zielinski 2002).

Therefore, characterizing aerosols not only requires describing their spatial and temporal distributions but their multi-component composition, particle size distribution and physical properties as well. Aerosols formed over land by either primary or secondary formation processes are transported over the oceans and contribute substantially to the aerosol concentrations over the oceans (Kastendeuch and Najjar 2003; Smirnov et al. 2003; Glantz et al. 2006). Sea spray aerosol is directly produced at the sea surface through the interactions between wind and surface waves

A. Strzalkowska (✉) · T. Zielinski · P. Makuch · P. Pakszys · T. Petelski
Institute of Oceanology of Polish Academy of Sciences, Sopot, Poland
e-mail: strzalkowska@iopan.gda.pl

T. Zielinski et al. (eds.), *Impact of Climate Changes on Marine Environments*,
GeoPlanet: Earth and Planetary Sciences, DOI 10.1007/978-3-319-14283-8_2

(Hobbs 2000). Sea salt aerosols are among most abundant components of atmospheric aerosols, and thus they exert strong influence on radiation, cloud formation, meteorology and chemistry of the marine atmosphere. An accurate understanding and description of these mechanisms is crucial to modeling climate and climate change (Smirnov et al. 2009, 2011). Secondary aerosol formation from gases released from the sea surface and even ship emissions contribute significantly to marine atmosphere aerosol loading.

It has been reported that the mass concentrations from sea spray aerosol and desert dust show the largest aerosol contributions on a global scale (Andreae and Rosenfeld 2008; Jickells et al. 2005; Lewis and Schwartz 2004; de Leeuw et al. 2011; Petelski et al. 2014). Aerosol particles are important both because they affect atmospheric processes and, in case of the world's oceans, after deposition to the sea surface, because they affect processes in sea water.

Aerosols have a strong impact on climate both due to scattering and absorption of incoming solar radiation (direct effect) and through their effects on cloud properties and associated cloud albedo (first indirect effect) and precipitation (second indirect effect). The appropriate correction of the atmospheric impact on the registered signal is an important problem in the remote sensing of the Earth's surface, and it is especially significant in areas, such as the Baltic Sea basin, which are very urbanized and industrialized. A thorough understanding and explanation of aerosol impact on light transmission in the atmosphere requires knowledge of aerosol optical properties, such as extinction, phase function and single scattering albedo, as well as microphysical aerosol properties, such as size distribution and light refractive index. This is especially relevant concerning knowledge of the real and imaginary parts of the light refractive index on aerosol particles including mineral dust additions (de Leeuw et al. 2011; Zielinski and Zielinski 2002). The optical properties of dust particles are important in calculations of the solar radiation that reaches the Earth's surface, and they force climatic changes in the areas where their concentrations are high, e.g. Baltic Sea, a typical regional sea, surrounded by highly industrialized areas (Dzierzbicka-Głowacka et al. 2013).

The ground-based methods are, in principle, the easiest to use and the most accurate monitoring systems. Aerosol optical depth (AOD) is the single most comprehensive variable to remotely assess the aerosol burden in the atmosphere using ground-based instruments. Knowledge of the real variations of this parameter facilitates the solution of problems with solar radiation transmission through the atmosphere as well as those concerned with climatology and remote sensing of the seas and oceans. Therefore, the AOD is used in local studies on aerosols, their role in atmospheric pollution and to make atmospheric corrections to satellite remotely sensed data (Zielinski and Zielinski 2002).

Sea salt is the most characteristic type of marine aerosol, and it enters the atmosphere due to the strong influence of wind on the sea surface. As a result, wind waves are formed and the particles are precipitated from the wave crests. Emission of sea salt to the atmosphere strongly depends on the force, speed and direction of wind (Jacob et al. 1995). Marine aerosols also can be generated by rainfalls or acoustic waves (Blanchard and Woodcock 1957; Blanchard 1963; Fitzgerald 1991).

In this paper we discuss the results of the studies of aerosol optical properties under the conditions of breaking waves in the coastal area of the Polish coast on Baltic Sea. We show that the sea salt production in the coastal area significantly changes the optical properties of aerosols, hence influencing the radiative budget on a local scale.

2 Methodology

The studies took place at the Coastal Research Station (CRS) in Lubiatowo on the Polish Baltic coast (54°48′42.0″N, 017°50′25.6″E), which is located about 70 km northwest of Gdansk (Fig. 1) on a typical South Baltic sandy coast. The area is dominated by westerly and south-westerly winds, which are strongest in the autumn and winter. Water salinity amounts to 7.5 PSU during summer and 7.7 PSU in the winter (http://mlb.ibwpan.gda.pl/index.php/en/).

The shore in the vicinity of the CRS Lubiatowo is an open and natural beach, characterized by a gentle slope (about 1.5 %). Due to existence of the sand bars, waves approaching the shore from deep sea are subject to significant transformation in the surf zone and most of wave energy is dissipated due to multiple wave breaking (http://mlb.ibwpan.gda.pl/index.php/en/). Such conditions are very favorable for the production of sea salt particles.

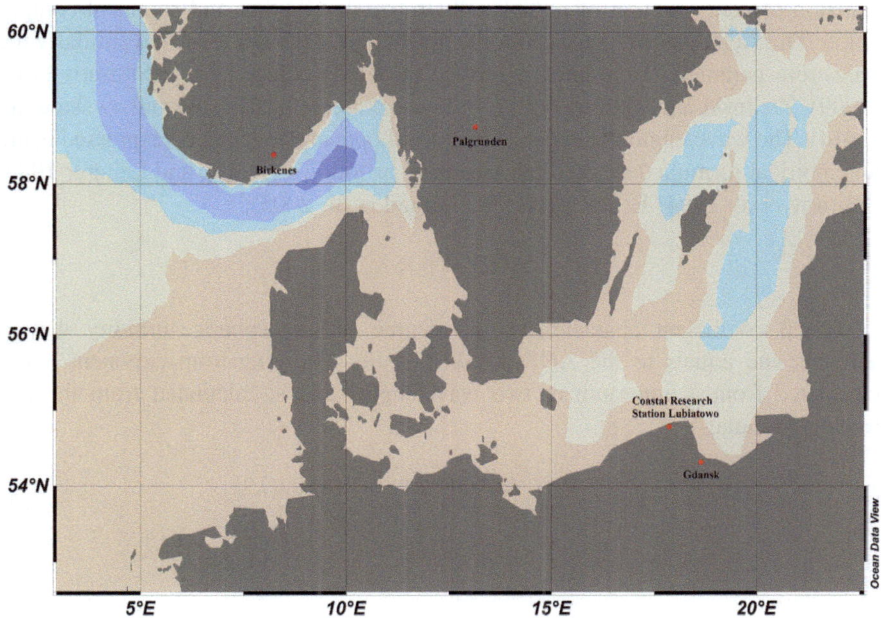

Fig. 1 Location of the coastal research station in Lubiatowo

The experiment lasted between 17 and 27 July 2012. The research team measured the aerosol optical depth (AOD) using a Microtops II sun photometer each measurement day started early in the morning and lasted till late afternoon weather permitted (no clouds or rain). Additionally, information on meteorological data (temperature, pressure, relative humidity) were collected from the appropriate weather services (e.g. www.rp5.kz) and the air mass back trajectories have been calculated using HYSPLIT (Draxler and Rolph 2014). Information on the AOD and types of aerosol particles were taken from the AERONET Birkenes station in southern Norway. Finally, the MODIS images were used to verify the AOD for the studied area.

The Microtops II sun photometer (produced by Solar Light Company USA) is a hand-held device capable of measuring aerosol optical depth of atmospheric aerosols, direct irradiance in each band and the water vapor column. The device is equipped with five channels and each of channels has different wavelength value (380, 440, 500, 675, 870 nm). The measurement procedure results in quick, lasting only 10 s, scanning of vertical column of atmosphere. Microtops II measurements were made in a series of five "shots". The scans are completed one after another and group into series for two minute time intervals. Only the lowest value is used since such value guarantees the least disturbed measurement of all 5. The full information on the Microtops II sun photometer and the Langley calibration technique used by the authors have been described in Morys et al. (2001), Zielinski et al. (2012), Strzalkowska et al. (2014).

The aerosol optical depth, a dimensionless, wavelength dependent parameter which refers to the weakening of direct sunlight passing through the atmosphere is a function of concentration of particles, their size distribution and chemical composition. The AOD value changes with the height above the sea level and similarly the Ängström parameter have been described in details before by e.g. Smirnov et al. (2009), Zielinski and Zielinski (2002), Zawadzka et al. (2014), Strzalkowska et al. (2014). The wavelength dependence of aerosol optical depth can be expressed using an empirical formula described by Ängström as follows (Weller and Leiterer 1988; Smirnov et al. 1994; Eck et al. 1999; Carlund et al. 2005):

$$\mathrm{AOD} = \beta \times \lambda^{-\alpha} \tag{1}$$

The β coefficient characterizes the degree of atmospheric turbidity due to aerosols and equals to the AOD for $\lambda = 1\ \mu\text{m}$. The Ängström exponent (α) is calculated from a minimum of two wavelengths and is calculated from the following formula:

$$\alpha(\lambda_1, \lambda_2) = \frac{\ln \mathrm{AOD}(\lambda 1) - \ln \mathrm{AOD}(\lambda 2)}{\ln \lambda 1 - \ln \lambda 2} \tag{2}$$

3 Results and Discussion

The studies lasted between 17 and 27 July 2012. During this period good photometric conditions, i.e. with no or small number of scattered clouds were observed only on 17, 18, 20 and 21 July. Therefore, in the reminder of the paper only these days have been further analyzed. The wind and sea wave conditions at the study site on these days are presented in Table 1.

During these days the air masses arrived from the westerly sectors. The air mass back trajectories calculated using the NOAA HYSPLIT Model are presented in Fig. 2.

It is clear that during all days the air masses were advected over the CRS in Lubiatowo from the north-westerly directions at all three altitudes above the sea level (500, 1,500 and 3,000 m). Wind speeds were relatively high, especially the wind gusts were at the same level or even slightly increasing. These conditions lasted for 5 days resulting in waves of heights around 2 m or even higher, which were breaking in the area, creating a wide breaking zone (over 500 m wide). In such conditions sea salt production is enhanced and this could be seen with the AOD values.

3.1 17 July 2012

On 17 July 2012 a large storm cloud system developed over the northern Poland, which resulted in mild rains lasting until noon. Thus on that day the Microtops II measurements at the CRS station in Lubiatowo were made every 30 min after 14:15 UTC. The average daily AOD at 500 nm was up to 0.154 and Ängström exponent of about 0.8 (Fig. 3).

Analyses of the MODIS (The Moderate Resolution Imaging Spectroradiometer) (not shown) pictures from both the Terra and Aqua instruments reveal large cloud systems over and the AOD at 500 nm not exceeding 0.2.

Table 1 Wind and sea wave conditions at the station in Lubiatowo on 4 selected measurement days

Date	Wind speed [min–max] (m/s)	Wind gust [min–max] (m/s)	Significant wave height (m)	Maximum wave height (m)
17.07	1–7	2–16	1–1.5	1.5–2
18.07	1–6	2–16	1–1.8	1.5–3
20.07	2–8	4–18	1–2	1.5–3
21.07	2–6	4–14	1–1.5	1.5–2.2

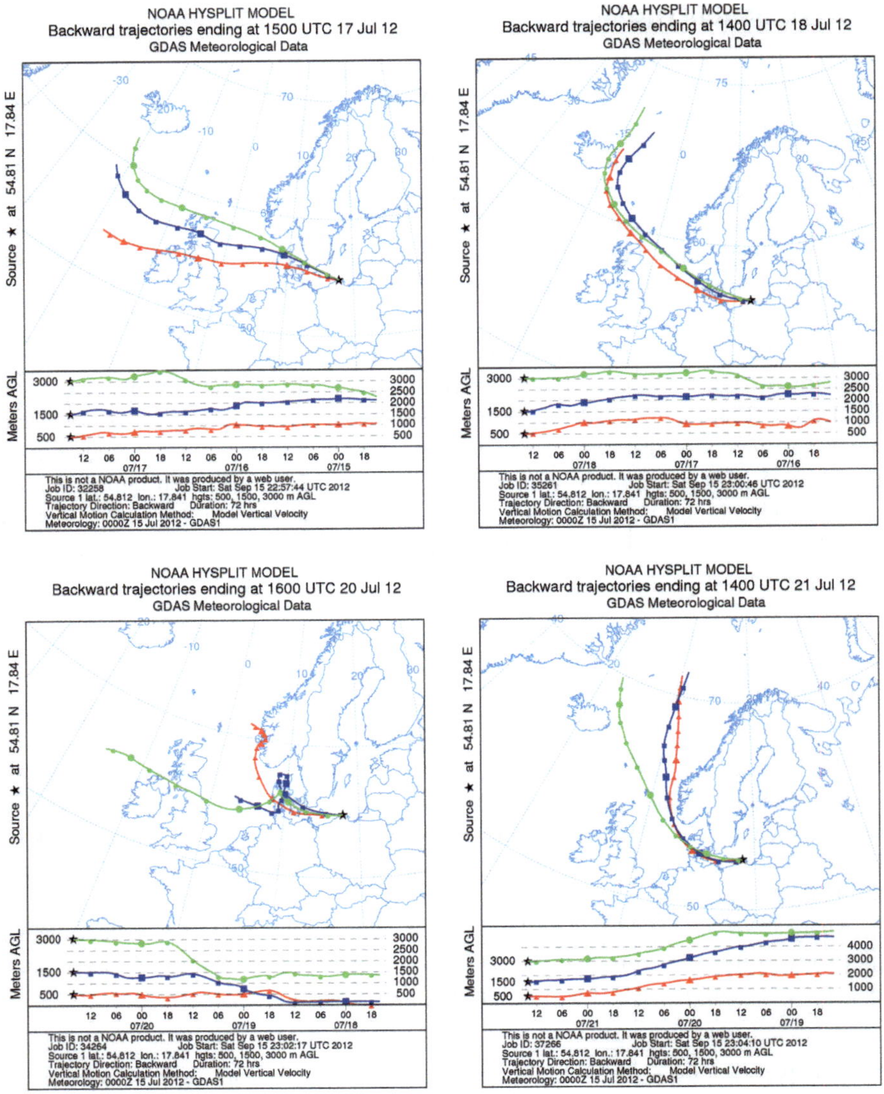

Fig. 2 Air mass back trajectories calculated for 17, 18, 20 and 21 July 2012 for Lubiatowo using the NOAA HYSPLIT model

The back trajectories obtained from the NOAA HYSPLIT Trajectory Model show the advection of air masses over the CRS station in Lubiatowo from north-west. Prior to the study area the air masses passed over the AERONET station in Birkenes (Norway). With an average daily wind speed of 20 km/h on 17 July and a distance of 700 km between the CRS station and the Birkenes station the same air

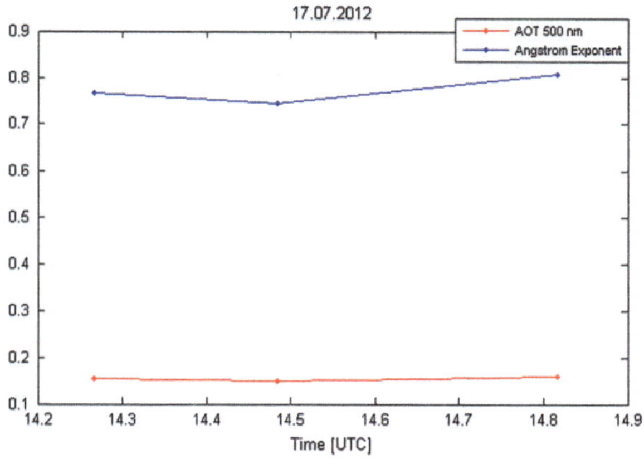

Fig. 3 The AOD (500 nm) and Ängström exponent (440–870 nm) on 17 July 2012 at the CRS station in Lubiatowo

masses passed the AERONET station some 35 h before the CRS station. The average daily AOD at 500 nm (Level 2) on 16 July in Birkenes was 0.053 and the Angstrom exponent (440–870 nm) equaled to 1.469, indicating presence of rather small particles in the air (Fig. 4a, b). This observation is also supported by a dominating role of fine mode over the coarse particles measured in Birkenes.

The comparison of the AERONET and the Lubiatowo data indicates the impact of breaking waves on aerosol production in Lubiatowo—AOD is an order of magnitude higher than in Birkenes and the Angstrom exponent indicates the presence of coarse mode particles, due to the breaking wave production of fresh particles, namely sea salt.

3.2 18 July 2012

On 18 July entire north Poland was under the influence of Atlantic air masses. A warm front from over Germany arrived over the study area. On that day Microtops II based measurements were made between 8:00 (UTC) until 13:15 (UTC). An average daily AOD value at 500 nm above the CRS station was 0.16 and the Ängström exponent was 0.65 (Fig. 5).

Again, during the entire day thick, scattered clouds were observed, which limited the number of sun photometric measurements. The cloud cover is evident from the very poor number of data presented with the MODIS (not shown in this article). The Aqua instrument registered only clouds, while the Terra instrument showed the AOD values not exceeding 0.2.

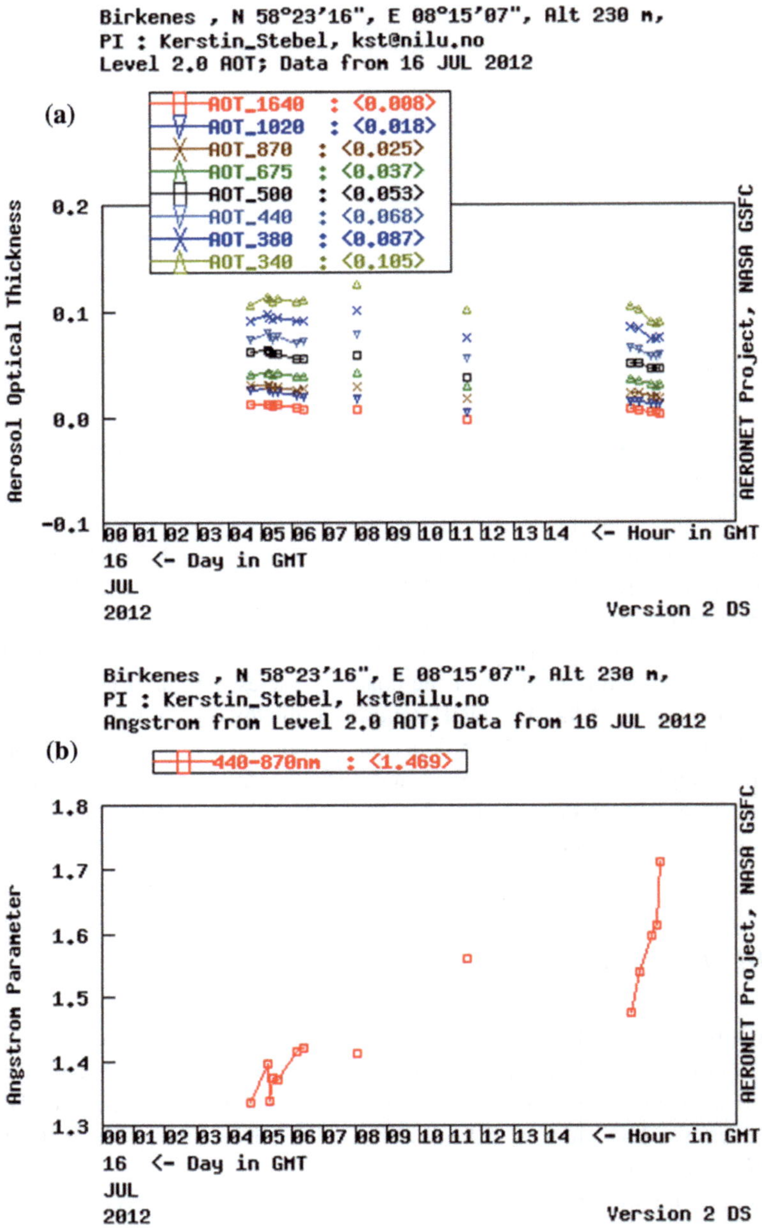

Fig. 4 The level 2.0 AOD (**a**) and the Ängström exponent (440–870 nm) (**b**) at the AERONET Birkenes station on 16 July 2012

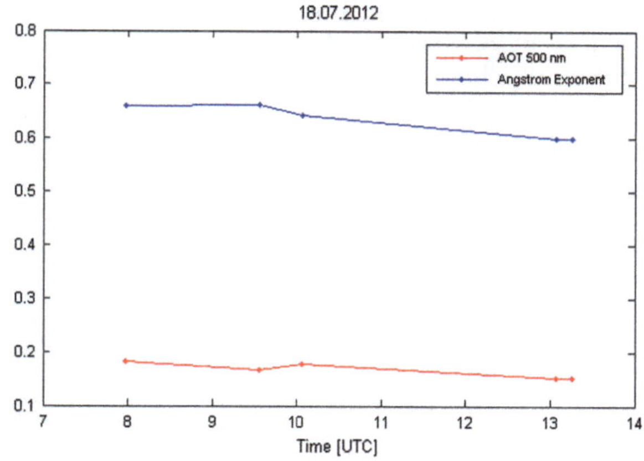

Fig. 5 The AOD (500 nm) and Ängström exponent (440–870 nm) on 18 July 2012 at the CRS station in Lubiatowo

NOAA HYSPLIT Trajectory Model shows that the air masses were advected to the CRS station from the west. The air masses passed two AERONET station on the way, the Birkenes station and the Palgrunden station (on 16 July the average AOD was 0.068 at 555 nm). Similar to 17 July 2012 the average daily wind speed of 20 km/h on 18 July the same air masses passed the AERONET station in Birkenes some 35 h before the CRS station. The AOD and Ängström exponent measured at the Birkenes station on 17 July are presented in Fig. 6a, b.

The average AOD at 500 nm at the Birkenes station was 0.050 and the Ängström exponent—1.803. These values indicate clear air and with presence of rather small particles. This situation is confirmed by the fine (0.045) to coarse ratio (0.006) for that day. Again these values are one order of magnitude lower than those obtained above the CRS station in Lubiatowo, 0.16 and 0.65, respectively. Due to wind speed and height of breaking waves these values indicate the presence of sea salt particles in the study area.

3.3 20 July 2012

On 20 July active atmospheric fronts caused storms and rains over the northern Poland. The Arctic air masses were blocked from being advected over the study area. Low pressure system caused strong winds. On 20 July the Microtops II measurements were made between 12:40 (UTC) and 15:05 (UTC). The daily average AOD at 500 nm was 0.179 and the Ängström exponent—0.74 (Fig. 7).

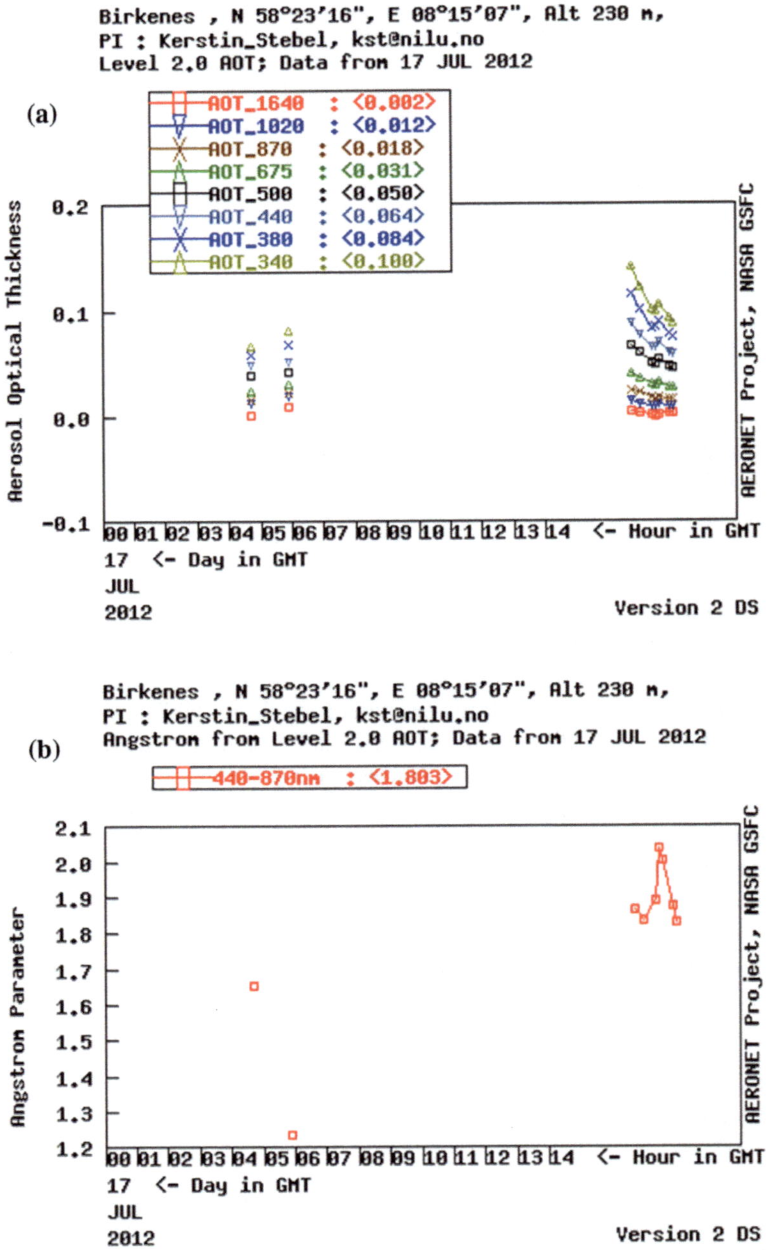

Fig. 6 The level 2.0 AOD (**a**) and the Ängström exponent (440–870 nm) (**b**) at the AERONET Birkenes station on 17 July 2012

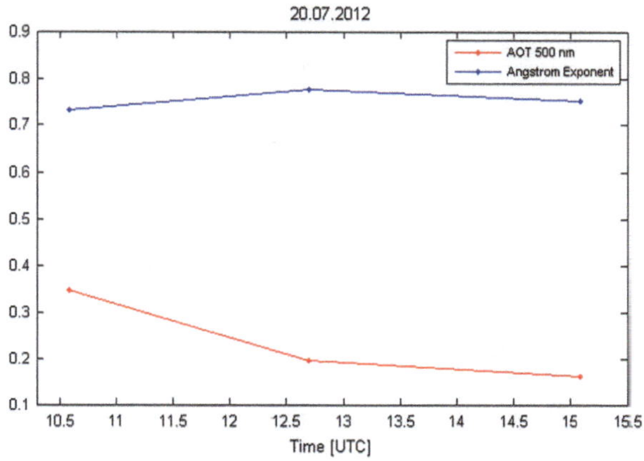

Fig. 7 The AOD (500 nm) and Angstrom exponent (440–870 nm) on 20 July 2012 at the CRS station in Lubiatowo

The analyses of the AOD from the MODIS were impossible due to the cloud coverage over the area.

Again the analyses of the back trajectories reveal that the air masses were advected to the study area from the west and they passed the Birkenes station. With the stable wind speed conditions the air masses passed the Birkenes station on 18/19 July (Fig. 8).

The average AOD at 500 nm at the Birkenes station was 0.043 and the Angstrom exponent—1.918, which indicate clear air and with presence of rather small particles. This situation is confirmed by the fine (0.042) to coarse mode ratio (0.005) for that day. These values are one order of magnitude lower than those obtained above the CRS station in Lubiatowo, 0.179 and 0.74, respectively, which indicate presence of sea salt particles due to high breaking waves

3.4 21 July 2012

On 21 July 2012 the low pressure system moved towards north (Scandinavia) and the northern Poland was under the cold front. On that day the Microtops II measurements were made between 12:03 (UTC) and 13:46 (UTC). The average AOD at 500 nm was 0.2 and the Ängström exponent—0.58 (Fig. 9).

The analyses of the MODIS pictures showed that the Terra instrument registered clouds and the Aqua instrument provided AOD at 500 nm at a level of 0.2 for the study area.

Just like on previous days the air masses were advected to the study area from the west and they passed the Birkenes station. With the stable wind speed conditions the air masses passed the Birkenes station on 19/20 July. The daily average AOD at 500 nm on 20 July was 0.038 and the Ängström exponent—1.814 (Fig. 10).

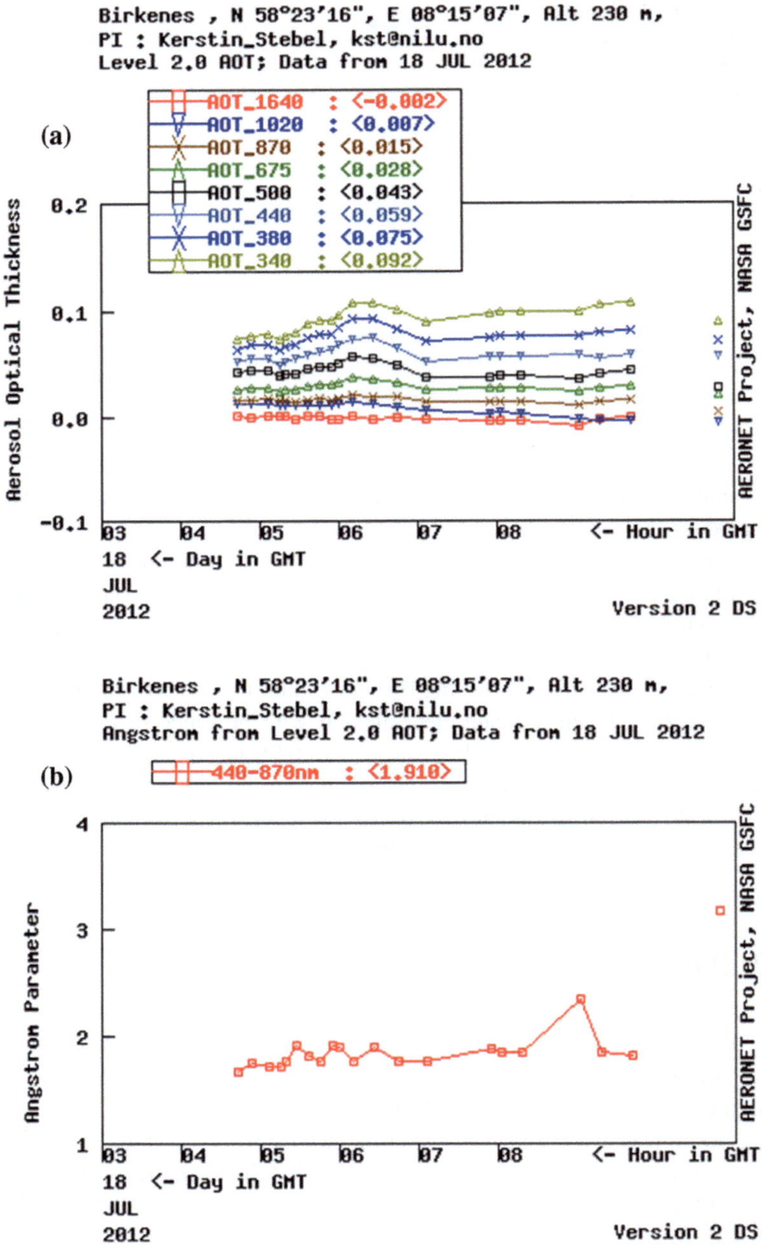

Fig. 8 The level 2.0 AOD (**a**) and the Angstrom exponent (440–870 nm) (**b**) at the AERONET Birkenes station on 18 July 2012

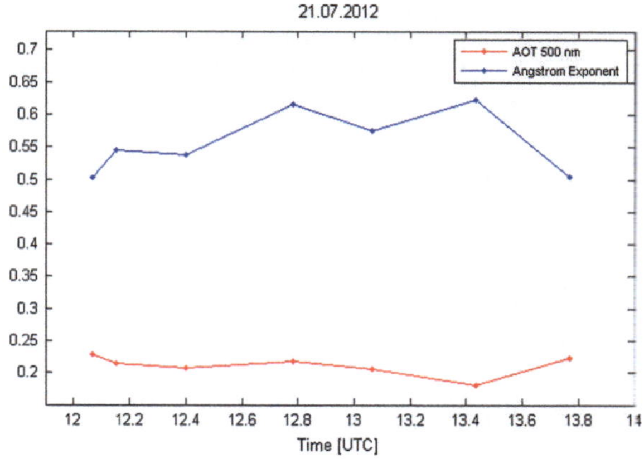

Fig. 9 The AOD (500 nm) and Ängström exponent (440–870 nm) on 21 July 2012 at the CRS station in Lubiatowo

Similar to the previous measurement days the values of the AOD and the Ängström exponent are an order of magnitude lower in Birkenes than in Lubiatowo. The fine (0.033) to coarse mode ratio (0.007) for that day shows the presence of small particles in the air, which is again in total opposition to the situation at the CRS study station.

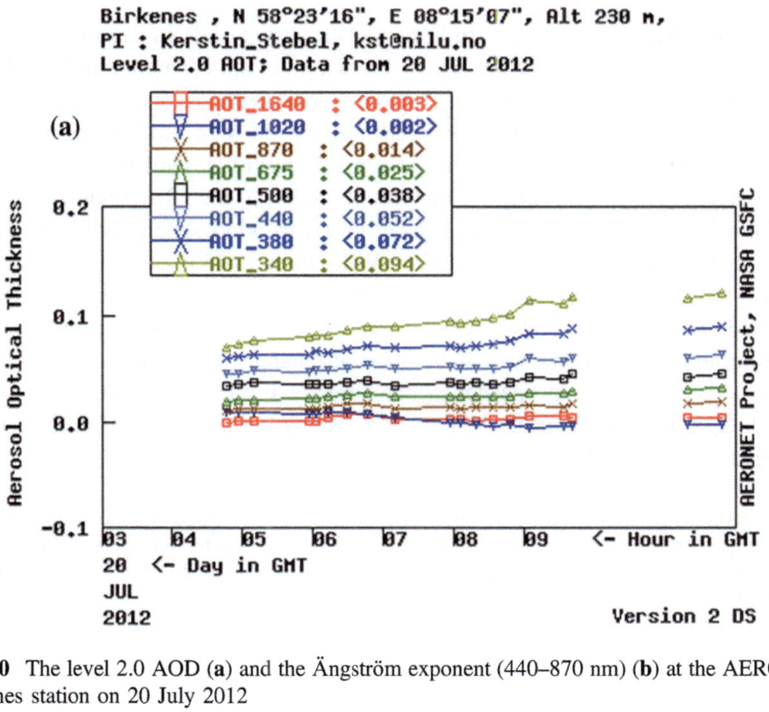

Fig. 10 The level 2.0 AOD (**a**) and the Ängström exponent (440–870 nm) (**b**) at the AERONET Birkenes station on 20 July 2012

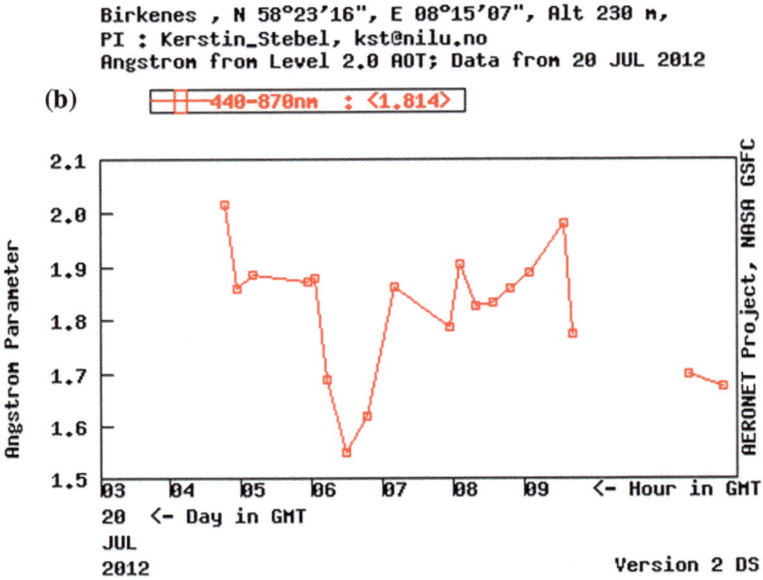

Fig. 10 (continued)

4 Conclusions

The results obtained from our studies indicate strong shift in aerosol size distribution due to the impact of the wave breaking. Despite small number of measurement points due to adverse weather conditions (cloudiness) with very advantageous wind directions and speeds during the five measurement days we were able to show a local change in the aerosol ensemble. During all studied days we registered AOD of an order of magnitude higher at the CRS station than at the AERONET station in Birkenes a day before (with wind speed and direction this should have been the same air mass). With high breaking waves and a large breaking zone (500 m) local production of sea salt must have been significantly increased. This is confirmed by a relatively low Ängström values around 0.6–0.7, versus around 1.7 in Birkenes. Such situation changes the local aerosol situation and thus their radiative properties which directly influence the radiative forcing. Even though we show a local phenomenon, it may have a more spatial impact and this needs further studies.

Acknowledgments The support for this study was partly provided by the project Satellite Monitoring of the Baltic Sea Environment—SatBałtyk founded by European Union through European Regional Development Fund contract No. POIG 01.01.02-22-011/09. It has also been made within the framework of the NASA/AERONET Program.

References

Andreae MO, Rosenfeld D (2008) Aerosol-cloud-precipitation interactions. Part 1: The nature and sources of cloud-active aerosols. Earth Sci Rev 89:13–41

Blanchard DC (1963) The electrification of the atmosphere by particle from bubbles in the sea. Prog Oceanogr 1:171–202

Blanchard DC, Woodcock AH (1957) Bubble formation and modification in the sea and its meteorological significance. Tellus 9:145–158

Carlund T, Hakansson B, Land P (2005) Aerosol optical depth over the Baltic Sea derived from AERONET and SeaWiFS measurement. Int J Remote Sens 26(2):233–245

de Leeuw G, Kinne S, Leon JF, Pelon J, Rosenfeld D, Schaap M, Veefkind PJ, Veihelmann B, Winker DM, von Hoyningen-Huene W (2011) Retrieval of aerosol properties. In: Burrows JP, Platt U, Borrell P (eds) The remote sensing of tropospheric composition from space. Springer, Berlin, pp 359–313. ISBN: 978-3-642-14790-6. doi:10.1007/978-3-642-14791-3 (536 pp)

Draxler RR, Rolph GD (2014) HYSPLIT (hybrid single-particle lagrangian integrated trajectory) model access via NOAA ARL READY website (http://ready.arl.noaa.gov/HYSPLIT.php). NOAA Air Resources Laboratory, Silver Spring

Dzierzbicka-Głowacka L, Jakacki J, Nowicki A, Janecki M (2013) Activation of the operational ecohydrodynamic model (3D CEMBS)—the hydrodynamic part. Oceanologia 55(3):519–541. doi:10.5697/oc.55-3.519

Eck TF, Holben BN, Reid JS, Dubovik O, Smirnov A, O'Neill NT, Slutsker I, Kinne S (1999) Wavelength dependence of the optical depth of biomass burning, urban, and desert dust aerosols. J Geophys Res 104(D24):31, 333–31, 349. doi:10.1029/1999JD900923

Fitzgerald JW (1991) Marine aerosols: a review. Atmos Environ 25A:533–545

Glantz P, Nilsson DE, von Hoyningen-Huene W (2006) Estimating a relationship between aerosol optical thickness and surface wind speed over the ocean. Atmos Chem Phys Discuss 6:11621–11651

Hobbs PV (2000) Introduction to atmospheric chemistry. Cambridge University Press, Cambridge, pp 82–100

Jacob DJ, Andreas MO, Bigg EK, Duce RA, Fung I, Hidy GM, Legrand M, Prospero JM, Raes F, Warren SG, Wiedensohler A (1995) Group report: what factors influence atmospheric aerosols, how have they changed in the past, and how might they change in the future. In: Charlson RJ, Heintzenburg J (eds) Aerosol forcing of climate. Wiley, New York, pp 183–195

Jickells TD, An ZS, Andersen KK, Baker AR, Bergametti G, Brooks N, Cao JJ, Boyd PW, Duce RA, Hunter KA, Kawahata H, Kubilay N, laRoche J, Liss PS, Mahowald N, Prospero JM, Ridgwell AJ, Tegen I, Torres R (2005) Global iron connections between desert dust, ocean biogeochemistry, and climate. Science 308:67–71

Kastendeuch PP, Najjar G (2003) Upper-air wind profiles investigation for tropospheric circulation study. Theor Appl Climatol 75:149–165

Lewis ER, Schwartz SE (2004) Sea salt aerosol production: mechanisms, methods, measurements, and models—a critical review. Geophysical monograph series, vol 152. American Geophysical Union, Washington, DC, p 413

Morys M, Mims FM III, Hagerup S, Anderson SE, Baker A, Kia J, Walkup T (2001) Design, calibration, and performance of MICROTOPS II handheld ozone monitor and Sun photometer. J Geophys Res 106(D13):14573–14582. doi:10.1029/2001JD900103

Petelski T, Markuszewski P, Makuch P, Jankowski A, Rozwadowska A (2014) Studies of vertical coarse aerosol fluxes in the boundary layer over the Baltic Sea. Oceanologia 56(4):697–710. doi:10.5697/oc.56-4.697

Smirnov A, Royer A, O'Neill NT, Tarussov A (1994) A study of the link between synoptic air mass type and atmospheric optical parameters, J Geophys Res 99(D10):20,967–20,982

Smirnov A, Holben BN, Eck TF, Dubovik O, Slutsker I (2003) Effect of wind speed on columnar aerosol optical properties at Midway Island. J Geophys Res 108(D24):4802 pp

Smirnov A, Holben BN, Slutsker I, Giles DM, McClain CR, Eck TF, Sakerin SM, Macke A, Croot P, Zibordi G, Quinn PK, Sciare J, Kinne S, Harvey M, Smyth TJ, Piketh S, Zielinski T, Proshutinsky A, Goes JI, Nelson NB, Larouche P, Radionov VF, Goloub P, Krishna Moorthy K, Matarrese R, Robertson EJ, Jourdin F (2009) Maritime aerosol network as a component of aerosol robotic network. J Geophys Res 114:1–10. doi:10.1029/2008JD011257

Smirnov A, Holben BN, Giles DM, Slutsker I, O'Neill NT, Eck TF, Macke A, Croot P, Courcoux Y, Sakerin SM, Smyth TJ, Zielinski T, Zibordi G, Goes JI, Harvey MJ, Quinn PK, Nelson NB, Radionov VF, Duarte CM, Losno R, Sciare J, Voss KJ, Kinne S, Nalli NR, Joseph E, Krishna Moorthy K, Covert DS, Gulev SK, Milinevsky G, Larouche P, Belanger S, Horne E, Chin M, Remer LA, Kahn RA, Reid JS, Schulz M, Heald CL, Zhang J, Lapina K, Kleidman RG, Griesfeller J, Gaitley BJ, Tan Q, Diehl TL (2011) Maritime aerosol network as a component of AERONET—first results and comparison with global aerosol models and satellite retrievals. Atmos Meas Tech 4:583–597. doi:10.5194/amt-4-583-2011

Strzalkowska A, Makuch P, Zawadzka O, Pakszys P (2014) A modern approach to aerosol studies over the Baltic Sea. Springer series—GeoPlanet: earth and planetary sciences. ISBN 978-3-319-03682-3

Weller M, Leiterer U (1988) Experimental data on spectral aerosol optical thickness and its global distribution. Beitr Phys Atmos 61(1):1–9

Zawadzka O, Makuch P, Markowicz KM, Zielinski T, Petelski T, Ulevicius V, Strzalkowska A, Rozwadowska A, Gutowska D (2014) Studies of aerosol optical depth with use of Microtops sun photometers and MODIS detectors in the coastal areas of the Baltic Sea. Acta Geophys 62 (2):400–422. doi:10.2478/s11600-013-0182-5

Zielinski T, Zielinski A (2002) Aerosol extinction and optical thickness in the atmosphere over the Baltic Sea determined with lidar. J Aerosol Sci 33(6):47–61

Zielinski T, Petelski T, Makuch P, Strzalkowska A, Ponczkowska A, Markowicz KM, Chourdakis G, Georgoussis G, Kratzer S (2012) Sudies of aerosols advected to coastal areas with use of remote techniques. Acta Geophys 60(5):1359–1385. doi:10.2478/s11600-011-0075-4

Annual Changes of Aerosol Optical Depth and Ångström Exponent over Spitsbergen

P. Pakszys, T. Zielinski, K. Markowicz, T. Petelski, P. Makuch, J. Lisok, M. Chilinski, A. Rozwadowska, Ch. Ritter, R. Neuber, R. Udisti and M. Mazzola

Abstract In this work we present the annual changes of two major, climate related aerosol optical parameters measured at three Spitsbergen locations, Ny-Alesund, Longyearbyen and Hornsund over a period between 2000 and 2012. We discuss the changes of aerosol optical depth (AOD) at 500 nm and the Ångström exponent (AE) (440–870 nm) measured with use of different types of sun photometers. For the measurement data we adopted several data quality assurance techniques and the calibration of the instruments was taken into consideration. The results obtained show that marine source has been a dominating of aerosol sources over Spitsbergen. Some years (2005, 2006, 2008 and 2011) show very high values of AOD due to strong aerosol events such as the Arctic Haze. In general the mean AOD values increase over the period of 2000 and 2012 over Spitsbergen. This may indicate the presence of larger scale of atmospheric pollution in the region.

Keywords Atmospheric aerosols · Aerosol optical properties · Sun photometry · Spitsbergen · iAREA · GAME

P. Pakszys (✉) · T. Zielinski · T. Petelski · P. Makuch · A. Rozwadowska
Institute of Oceanology, Polish Academy of Sciences, Sopot, Poland
e-mail: pakszys@iopan.gda.pl

K. Markowicz · J. Lisok · M. Chilinski
University of Warsaw, Warsaw, Poland

T. Zielinski
Centre for Polar Studies National Leading Research Centre, 60 Bedzinska Street, 41-200 Sosnowiec, Poland

Ch.Ritter · R. Neuber
Alfred-Wegener-Institute Helmholtz-Centre for Polar- and Marine Research, Potsdam, Germany

R. Udisti
University of Florence, Florence, Italy

M. Mazzola
Institute of Atmospheric Sciences and Climate, National Research Council, Bologna, Italy

T. Zielinski et al. (eds.), *Impact of Climate Changes on Marine Environments*,
GeoPlanet: Earth and Planetary Sciences, DOI 10.1007/978-3-319-14283-8_3

1 Introduction

The Arctic region is especially sensitive to climate change and its climate is modulated, in part, by atmospheric aerosols that affect the distribution of radiative energy passing through the atmosphere (Rozwadowska and Górecka 2012). Aerosols affect the surface-atmosphere radiation balance directly through interactions with solar and terrestrial radiation and indirectly through interactions with cloud particles. In Polar regions, where the surface albedo can exceed 0.85 (in VIS) in snow and ice covered areas, aerosols may cause significant warming at the ground (Tomasi et al. 2007; Engvall et al. 2008). While such effects are due mainly to the direct scattering and absorption of incoming solar radiation, exchanges of thermal radiation between the surface and the atmosphere enhance heating below aerosol layers (Stohl 2006; Fisher et al. 2010).

Atmospheric aerosols originate from a wide variety of sources in both marine and continental environments and their content varies significantly depending upon the air mass source and history (Petelski et al. 2014). These species are, in general, poorly accounted for in climate models. Better quantification of the radiative forcing by different types of aerosol is needed to improve predictions of future climate (Brock et al. 2011).

During the last century the temperature increase in the Arctic has been observed to be larger than the global average (IPCC 2013). The reason for this "Arctic amplification" relates to both the complex feedbacks that are active in the Arctic environment as well as the overall environmental conditions that are characteristic of the Arctic environment (Quinn et al. 2007). This increased warming results in positive feedback which further impacts the radiative balance via reduced surface albedo (Hudson 2011). Future changes in the Arctic are projected to progress rapidly and the projections show that the Arctic Ocean may be seasonally ice free in the next several decades. This will result in a more pronounced impact on atmospheric aerosol sources and sinks and on cloud properties and their distribution in the area (Petelski and Piskozub 2006).

Methods commonly used for monitoring atmospheric pollution (including aerosols) are optical ones, which collect data from a given point or a small area (Labow 1996; Dixon 1998; Drollette 2000; Smirnov et al. 2002). Studies using ground-based sun photometry are very effective in investigations of aerosol optical properties. Aerosol optical depth measured at different wavelengths is one of the key parameters in aerosol studies (Dubovik et al. 2002; Zielinski 2004; Markowicz et al. 2008; Mazzola et al. 2012; Zielinski et al. 2012). Also satellite remote sensing is a good approach to obtain the aerosol information over the Arctic region, for which appropriate aerosol models are required.

In this paper we describe the aerosol optical depth and Ångström exponent values measured at three locations in Spitsbergen. These stations include Hornsund in the south of the island, Longyearbyen in the center of the island and Ny-Alesund, in the north.

2 Site Characteristics and Instrumentation

2.1 Station Characteristics

The climate of Svalbard is dominated mostly by its northerly location, while the Norwegian Current and West Spitsbergen Current (which are a continuation of the North Atlantic Current) moderate its temperatures. The Arctic climate is the place where cold polar air from the north and west (high pressure over Greenland and the Polar basin) meets mild, wet sea air from the south (low pressure between Greenland and Spitsbergen) (Treffeisen et al. 2011; Rozwadowska et al. 2010). As a result very active cyclonic circulation and fronts with cloudy conditions, rain and strong winds are often reported in this region. These are major factors which determine the changeable weather over Svalbard, which in various parts of the archipelago is significantly different. The western part is warmer, while the interior has relatively more continental climate than the coasts.

Three sites in Spitsbergen are taken into consideration and they include all available AERONET (AErosol RObotic NETwork) data over Svalbard (http://aeronet.gsfc.nasa.gov/). These include stations in Hornsund (77°00′03″N, 15°33′36″E, at 10 m a.s.l), Longyearbyen (78°13′12″N, 15°38′56″E, at 30 m a.s.l.) and Ny-Alesund (78°55′44″N, 11°51′39″E, at 46 m a.s.l.) (Fig. 1).

Ny-Alesund is the highest above sea level and most northerly situated station located on Brøggerhalvøya and Kongsfjorden. The village is surrounded by

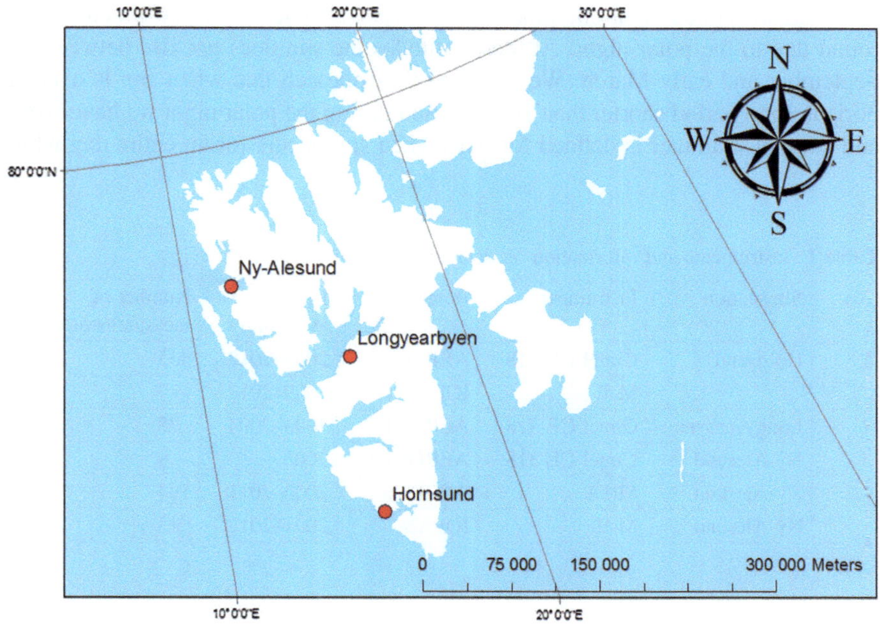

Fig. 1 Location of the research stations in Spitsbergen

mountains and tundra system. The Svalbard capital town of Longyearbyen—in the middle part of the island, is situated in the valley of Longyeardalen and on the shore of Adventfjorden. Hornsund is the southernmost fjord of the western side of Spitsbergen. The location of the stations has significant impact on the differentiation of air masses moving to the study area.

The AERONET network provides long-term, globally distributed observations of spectral aerosol optical depth (AOD) as well as Ångström exponent (AE). Ground-based remote sensing techniques are used to obtain long-term and continuous characterization of aerosols over the whole world.

In Ny-Alesund observations have also been performed in the AWIPEV (French-German Arctic Research Base at Koldeway station, Ny-Alesund) (http://www.awipev.eu/) (Herber et al. 2002). Since 2001 the Institute of Oceanology Polish Academy of Sciences (IOPAN) has performed their aerosol studies in the same area as the AERONET, i.e. in Longyearbyen, Hornsund and Ny-Alesund.

2.2 Instruments and Database Characteristics

The database used in this paper is composed of measurements performed in three different areas of Spitsbergen (Hornsund, Longyearbyen, Ny-Alesund) and different sun photometers as well as for different time intervals. The instrument and data information are provided in Table 1.

For all locations we use data from a period of March 2000–September 2012. At the Svalbard latitude sun photometric measurements cannot be performed all year round due to the polar night. At these latitudes the sun does not rise between late September and early March. We accepted the approach that when sun is over the horizon for a period shorter than 10 h/day and during the polar night we have winter and autumn. Summer is defined for days when sun occurs for an entire day while,

Table 1 Instruments and information on data availability

No.	Site/station	Instrument	Reference	Data availability	Number of measurement days
1	Hornsund	Cimel CE-318	AERONET	2005–2012	435
		M-II	IOPAN	2009–2012	9
2	Longyearbyen	Cimel CE-318	AERONET	2003–2004	78
3	Ny-Alesund	Cimel CE-318	AERONET	2006	9
	Ny-Alesund	SP1A	AWIPEV	2000–2011	594
	Ny-Alesund	M-II	IOPAN	2001–2012	13

spring—for more than 10 h a day and only for measurements which were carried out before the defined summer. Such an approach resulted in a reduced number of data. We use data from only spring and summer months. Secondly, only clear sky conditions enable to make any solar measurements and thus the number of "good" measurement days is also limited.

The AERONET protocols impose standardization of instruments, data quality, processing and calibration (Holben et al. 1998). The measurements are acquired with Cimel sun photometer CE-318. This automatic sun and sky radiometer has spectral interference filters centered at selected wavelengths: 340, 380, 440, 500, 670, 870, 1020 and 1640 nm. The real time operation of the data acquisition and motion steering are controlled by microprocessors. Sequence of the measurements is provided automatically every clear day, every 15 min (Holben et al. 1998). The data accuracy is 0.01 (visible solar radiation) or 0.02 (ultraviolet) (Smirnov et al. 2000). The AERONET provides tree levels of data: level 1.0 (raw data), level 1.5 (cloud-screened data) and level 2.0 (quality-assured data). For the detailed description see AERONET website (http://aeronet.gsfc.nasa.gov/).

The IOPAN based data obtained with a Microtops II sun photometer were collected and processed with the pre- and post-field calibration, automatic cloud clearing and were manually inspected. These portable instruments made by Solar Light Company are capable of measuring the total ozone column, total precipitable water vapor and aerosol optical depth (Morys et al. 2001; Ichoku et al. 2002). Each of these parameters is automatically derived by the instrument from equations installed by the manufacturer. The Microtops II instruments currently in use at the IOPAN have five channels, but they may have one of two configurations: 340, 440, 675, 870, 936 nm or 440, 500, 675, 870, and 936 nm. The estimates of uncertainty of the AOD in each channel oscillates around 0.02. Detailed instrument description is available at the AERONET webpage (http://aeronet.gsfc.nasa.gov) and has been presented by Markowicz et al. (2012) and Zawadzka et al. (2014).

In this paper, we also used the data from Koldeway Station in Ny-Alesund performed by AWIPEV. Data from 2000 until 2012 were obtained using a full-automatic sun photometer type SP1A produced by Dr. Schulz and Partner GmBH. The instrument covers a spectral range from 350 to 1,050 nm in 17 channels. It automatically tracks the sun, which ensures continuity of the measurements. The AOD uncertainty is 0.01 (Toledano et al. 2012).

3 Methodology

The AOD is a key parameter in aerosol studies which describes the entire atmospheric column and is derived from the Beer-Bouger-Lambert law. In case of Cimel and SP1A sun photometers the AOD is obtained using the following algorithm.

The total optical depth of the atmosphere (τ) is obtained from the absolute direct signal from the ground level (S(λ)):

$$S(\lambda) = S_0(\lambda) \cdot e^{(-\tau m)}, \tag{1}$$

where: $S_0(\lambda)$ is signal at the top of the atmosphere (with earth-sun distance correction), m—air mass. The AOD (τ_a) is obtained after subtraction of the Rayleigh optical depth (τ_R), contribution and ozone optical depth (τ_{O_3}) for the 670 nm channel:

$$\tau_a = \tau - \tau_R - \tau_{O_3}. \tag{2}$$

The Ångström Exponent (AE) is indicative of the size predominance. From spectral AOD at channels 440, 670 and 870 nm data are calculated AE:

$$\tau_a(\lambda) = \beta(\lambda)^{-\alpha}. \tag{3}$$

The final post-processing data including sequencing, cloud-screening is carried out with the AERONET protocols (Smirnov et al. 2000).

The Microtops II calculates the AOD value at each wavelength based on the channel's signal, its extraterrestrial constant, atmospheric pressure (for Rayleigh scattering), time and location. Solar distance correction is automatically applied. All optical depth calculations are based on the Bouguer-Lambert-Beer law. The AOD formula is as follows:

$$AOT_\lambda = \frac{ln(V_{0\lambda}) - ln(V_\lambda \cdot SDCORR)}{m} - \tau_R \cdot \frac{P}{P_0}, \tag{4}$$

where: $ln(V_{0\lambda})$ is the AOD calibration constant, V_λ is the signal intensity in (mV), SDCORR is the mean Earth-Sun distance correction, m is the optical air mass, τ_R is the Rayleigh optical depth, and P and P_0 are station pressure and standard sea-level pressure (1013.25 mB), respectively (Morys et al. 2001).

Typically, aerosol optical depths are derived from ground-based techniques. Sun photometer is a standard instrument which gives the integral for the total atmospheric column. This is the first step to build up the parameters which will determine the aerosol optical characteristics.

In our analyses we used Level 2.0 data. Such choice has already limited our data to those which have already been cloud-screened and quality assured. As a result we have obtained a total of 522 days and 11,387 measurements from all stations.

We present the AOD data only at a wavelength of 500 nm. We characterized the slope of these spectra characteristics by the Ångström Exponent, which is the function of the particle size distribution. It is calculated for the range 440–870 nm according to the AERONET protocol.

The presence of clouds is not always possible to detect, especially with thin Cirrus clouds or drifting snow crystals (Rozwadowska and Sobolewski 2010). Thus the data were also manually inspected with meteorological observation (WMO, MODIS) if necessary. This strategy was also followed with the Microtops II data.

For the IOPAN measurements we adopted a similar strategy to that of the AERONET. Data were collected during the IOPAN routine, annual expeditions (AREX-Arctic Expedition) or during dedicated campaigns within the scope of the research projects, such as e.g. iAREA (http://polandaod.pl/). IOPAN data were recalibrated with the strategy presented in the instrument User Guide and according to the Ichoku et al. (2002). Data were recalculated based on formula 4. Then the detection of clouds was checked with the satellite and the WMO. Data with presence of cirrus clouds were rejected. From each series of 5 shots only the lowest value was used. Sequencing the data into series of five "shots" with 2 min time limit allows to improve the quality of the data after choosing the best result.

The SP1A instruments operated at the AWIPEW station are calibrated in October at Zugspitze or in February at Izana/Tenerife, Spain using the well-known Langley procedure for solar applications (Shaw 1976). More details are available in Herber et al. (2002). Data from SP1A contain 594 days in 159,273 measurements. The AWIPEV data were cleared from instrument's error and a computational algorithm has been applied, in which we analyzed the 'suspect' data (errors, clouds, snowstorms, etc.). The 'suspect' data meet the following conditions:

(1) $AOD > 0.1$

In this case each AOD point which meets such criterion is classified as an event (haze, pollution, clouds etc.),

(2) $|AOD_2 - AOD_1| \geq 0.04$

In this case data when absolute value of the difference between successive measurements during the same day is higher or equals 0.04 has been chosen. This condition filters out Arctic Haze from the selected data. The expected variability of an Arctic Haze is very low during all analyzed events (AOD values are stable).

(3) $STD_{AOD} \geq 0.02$

Similar condition which informs about daily variability. Only the days which meet the previous conditions and with standard deviation higher or equal 0.02 have been left in this step.

After these 3 steps we were checking if the dates did not cover dates of data from other instruments. Only different dates have been left. The extracted data were evaluated with respect to Cloud—Aerosol Lidar and Pathfinder Satellite Observation (CALIPSO), MODIS data and also with the World Meteorological data for weather station in Ny-Alesund.

Similar conditions were adopted for the Angstrom exponent:

1. $AE \geq -0.2$ & $AE \leq 2$

In this case we selected the Angstrom exponent with extreme values, which could be a systematic error of instruments, especially for the SP1A for 4 years of measurements.

2. $|AE_2 - AE_1| \geq 2$

Difference between successive measurements as an absolute value during the same day higher or equals 2. Chosen values are very unusual during the same day with stable values of AOD.

3. $AE_{mean} - 2AEi \geq 2$

Difference between mean value of the AOD and double variable value higher than 2. This condition selects outliers from the mean AE value.

4 Results and Discussion

Looking more closely in the data we distinguished those for which AOD exceed 0.1. Those data were used for the classification of events:

(1) $AOD > MEAN_{AOD} + STD_{AOD}$ as an event
(2) $AOD > MEAN_{AOD} + 4STD_{AOD}$ as an extreme event

Such analyses allowed for the specification of the occurrence of 5.27 % of events in the entire data set. It gives a total of 8,326 out of 157,847 data rows. Extreme events account for 2.35 % of the data (3,720 cases), and most of them occurred in 2003, 2006, 2008 and 2010.

While AOD gives information about the aerosol loading, the AE is elated to aerosol size (type). The distribution of scatter plot enables to identify aerosol sources and size distribution (Toledano 2007). For that purpose we have prepared a scatter plot of AOD versus AE with information of number of measurements. Aerosols can be divided into the following types:

1. Marine aerosols

The pure marine should be located in the region with AOD < 0.15 and AE 0.5–1.7. This type is confined to the so-called accumulation mode and is present above the oceanic areas. The potential transport of continental aerosols over maritime environments interferences with this type of particles, which changes density and particle size distribution.

2. Continental and biomass burning aerosols

Continental origin aerosols are also expected at coastal sites. This type mainly consists of fine particles (<2.5 μm), and presents high values of AE (above 1). The AOD is very variable and depends on the weather conditions (mostly around 0.15–0.30), and that aerosols could be less or more polluted (AOD > 2.5 during e.g. forest fires etc.). Biomass burning aerosols are characterized by turbid atmosphere and large values of the AOD are reported.

3. Desert Dust aerosols

Particles are characterized by very turbid atmosphere. Very similar values of AOD result in low AE values. The AOD increase from 0.2–0.3 up to 1.2 against decreasing values of AE from 1 to 0.

4. Mixed types of aerosols

Coastal and marine produce a mixed type—with typical AOD < 0.15 and AE 0.3–0.6.

In Fig. 2 we present all AOD and AE data collected between 2000 and 2012 in all three Spitsbergen locations, Ny-Alesund, Longyearbyen and Hornsund altogether. The data have been collected in spring and summer seasons. In Fig. 3 we show a scatter plot of AOD (500 nm) versus AE for all data from the discussed stations.

The data show a natural temporal ordering which is related to seasonal changes of aerosol loads into the Svalbard region. The AOD decreases from the higher events during springs (mean for spring $\sim 0.085 \pm 0.046$) to more or less stable situation in summers (mean for summer $\sim 0.063 \pm 0.042$). There were no events during the years 2000–2002, in these years we have clear occurrence of marine aerosols. Mean for those years vary as follows: 0.084 ± 0.018, while the Ångström exponent: 1.269 ± 0.194.

Anthropogenic contamination advected with air masses from midlatitudes reach the polar regions seasonally, especially in early spring and summertime. In extreme cases of such advections or due to photochemical transformations of locally observed aerosols we deal with the so-called Arctic Haze. Several studies have

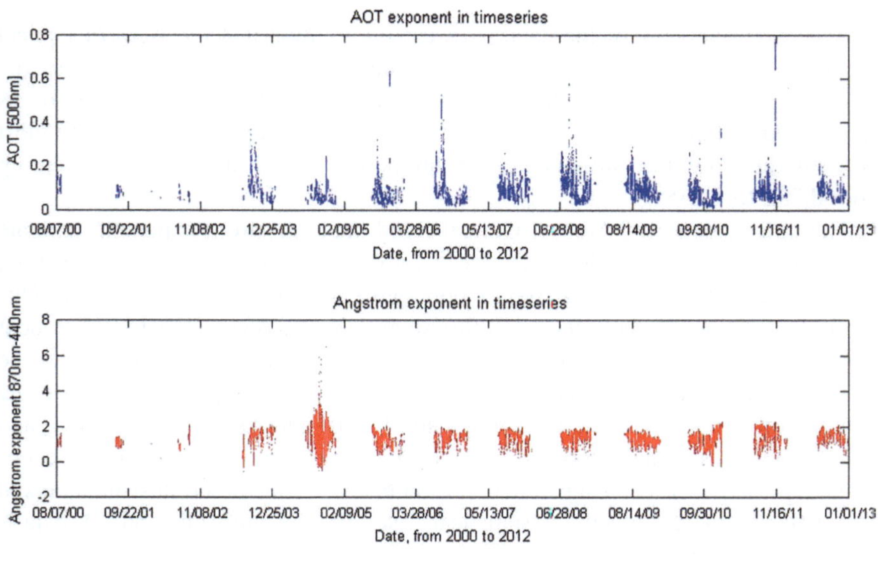

Fig. 2 AOD (500 nm) and AE for all three stations between 2000 and 2012

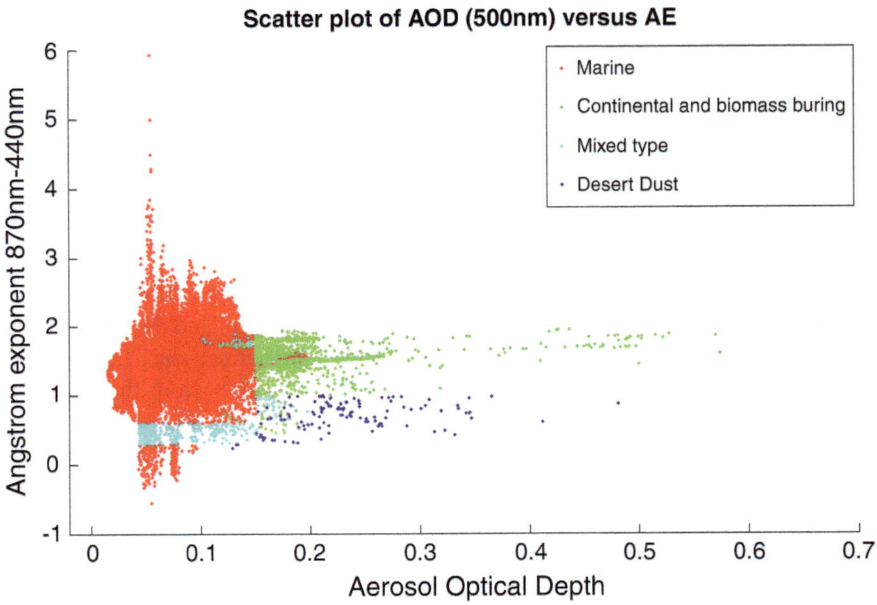

Fig. 3 Scatter plot of AOD (500 nm) versus AE for all data from the discussed stations

shown the occurrence of the phenomenon in 2005 and 2006 (Quin et al. 2007; Engvall et al. 2008; Rozwadowska et al. 2010). The occurrence of the strongest events during summer 2004, 2010 and spring 2006, 2008 are classified as extreme events. Those high AOD values have been related to the inflow of continental air masses.

A general overview of the data gives the information that the Aerosol Optical Depth is much higher these years than a decade before. With respect to the Ångström Exponent, which almost always follows the AOD values, most of its higher values should be related to the anthropogenic influence. Each year the most frequent value oscillates around 1.2–1.5, but between 2001 and 2011 they went up from 0.732 to 1.835.

The independence of Aerosol Optical Depth and Ångström exponent illustrate the origin of aerosols. While AOD gives an information about the aerosol loading, the AE is related to aerosol size (type), both make an interpretation of the data. This nonlinear relationship between variables shows that marine is the most frequent source that can be observed, presented in a form of the largest concentration on the left-side of the plot within an entire spectrum of Ångström exponent and the AOD changing from 0 to 0.15. The remaining part of AOD and AE spectra exceed 1, and this characterizes an anthropogenic source, such as e.g. biomass burning or continental type particles. The Ångström exponent below 1—the desert dust and a mixed type, between these two sets. A statistical description of the AOD and AE data is provided in Table 2.

Table 2 Statistics of AOD (500 nm) and the AE with time (between 2000 and 2012) and among stations

Station	Nr of pts	Year	AOD (500 nm)				Ångström exponent			
			Minimum	Maximum	Mean ± StD	Variance	Minimum	Maximum	Mean ± StD	Variance
Hornsund	853	2005	0.023	0.319	0.073 ± 0.042	0.002	0.370	1.793	1.148 ± 0.277	0.077
	923	2006	0.022	0.522	0.111 ± 0.099	0.010	0.208	1.951	1.326 ± 0.332	0.110
	1,004	2007	0.037	0.256	0.089 ± 0.032	0.001	0.300	1.894	1.269 ± 0.387	0.150
	1,913	2008	0.023	0.574	0.102 ± 0.055	0.003	0.211	1.943	1.456 ± 0.253	0.064
	1,430	2009	0.041	0.285	0.104 ± 0.039	0.002	0.418	1.806	1.378 ± 0.245	0.060
	401	2010	0.049	0.305	0.095 ± 0.035	0.001	0.499	1.877	1.234 ± 0.338	0.115
	1,534	2011	0.036	0.240	0.087 ± 0.030	0.001	0.235	1.977	1.371 ± 0.419	0.176
	1,336	2012	0.025	0.209	0.082 ± 0.032	0.001	0.391	2.105	1.561 ± 0.271	0.073
Long-year-byen	966	2003	0.029	0.365	0.095 ± 0.060	0.004	0.447	2.077	1.609 ± 0.261	0.068
	491	2004	0.024	0.114	0.048 ± 0.013	0.000	0.491	1.839	1.357 ± 0.251	0.063
Ny-Alesund	1,184	2000	0.073	0.195	0.110 ± 0.025	0.001	0.879	1.616	1.317 ± 0.156	0.024
	2,179	2001	0.053	0.110	0.077 ± 0.016	0.000	0.206	1.432	1.093 ± 0.148	0.022
	1,233	2002	0.037	0.115	0.066 ± 0.015	0.000	0.619	2.080	1.399 ± 0.279	0.078
	888	2003	0.044	0.139	0.067 ± 0.015	0.000	−0.185	2.164	0.755 ± 0.519	0.269
	10,515	2004	0.027	0.245	0.088 ± 0.046	0.002	−0.192	2.318	1.609 ± 0.485	0.235
	2,499	2005	0.016	0.633	0.106 ± 0.093	0.009	0.313	2.317	−1.334 ± 0.369	0.136
	7,039	2006	0.017	0.264	0.062 ± 0.035	0.001	0.535	1.802	1.493 ± 0.189	0.036
	18,442	2007	0.028	0.189	0.075 ± 0.032	0.001	1.025	1.937	1.580 ± 0.164	0.027
	16,239	2008	0.025	0.183	0.078 ± 0.038	0.001	1.046	1.875	1.536 ± 0.153	0.023
	29,862	2009	0.035	0.208	0.089 ± 0.032	0.001	0.773	1.995	1.360 ± 0.243	0.059
	28,935	2010	0.006	0.126	0.049 ± 0.021	0.000	−0.163	2.237	1.382 ± 0.319	0.102
	26,021	2011	0.035	0.162	0.067 ± 0.020	0.000	1.065	2.234	1.835 ± 0.147	0.022

The basic statistics, which are presented in the table were calculated for all three stations. The extremes (minimum and maximum) and central tendency (mean) were used to present the changing aerosol structure with years and among the different locations. The variance measures how the parameters are spread out each year and almost for whole data it tends to be very close to the mean (small values).

5 Conclusions

In this work we have discussed the changes of aerosol optical depth (AOD) at 500 nm and the Ångström exponent (AE) (440–870 nm) measured with use of different types of sun photometers. A general conclusion is that the results obtained over a period of 2000 and 2012 show that marine source has been a dominating of aerosol sources over Spitsbergen. Some years (2005, 2006, 2008 and 2011) show very high values of AOD due to strong aerosol events such as the Arctic Haze. In general the mean AOD values increase over the period of 2000 and 2012 over Spitsbergen. This may indicate the presence of larger scale of atmospheric pollution in the region. This conclusion has to be further verified by applying of chemical composition analyses.

Acknowledgments This research has been partly made within the framework of a Polish-Norwegian Research Programme operated by the National Centre for Research and Development under the Norwegian Financial Mechanism 2009–2014 in the frame of Project Contract No Pol-Nor/196911/38/2013.

References

Brock CA, Cozic J, Bahreini R, Froyd KD, Middlebrook AM, McComiskey A, Brioude J, Cooper OR, Stohl A, Aikin KC, de Gouw JA, Fahey DW, Ferrare RA, Gao R-S, Gore W, Holloway JS, H¨ubler G, Jefferson A, Lack DA, Lance S, Moore RH, Murphy DM, Nenes A, Novelli PC, Nowak JB, Ogren JA, Peischl J, Pierce RB, Pilewskie P, Quinn PK, Ryerson TB, Schmidt KS, Schwarz JP, Sodemann H, Spackman JR, Stark H, Thomson DS, Thornberry T, Veres P, Watts LA, Warneke C, Wollny AG (2011) Characteristics, sources, and transport of aerosols measured in spring 2008 during the aerosol, radiation, and cloud processes affecting Arctic Climate (ARCPAC) Project. Atmos Chem Phys 11:2423–2453. doi:10.5194/acp-11-2423-2011

Dixon GJ (1998) Laser radars produce three-dimensional pictures. Laser Focus World 4:129–136

Drollette D (2000) Ancient writings come to light. Photonics Spectra 4:40

Dubovik O, Holben B, Eck TF, Smirnov A, Kaufman YJ, King MD, Tanré D, Slutsker I (2002) Variability of absorption and optical properties of key aerosol types observed in worldwide locations. J Atmos Sci 59(3):590–608. doi:10.1175/1520-0469(2002)059<0590:VOAAOP>2.0.CO

Engvall A-C, Krejci R, Ström J, Treffeisen R, Scheele R, Hermansen O, Paatero J (2008) Changes in aerosol properties during spring-summer period in the Arctic troposphere. Atmos Chem Phys 8:445–462. http://www.atmos-chem-phys.net/8/445/2008/

Fisher JA, Jacob DJ, Purdy MT, Kopacz M, Le Sager P, Carouge C, Holmes CD, Yantosca RM, Batchelor RL, Strong K, Diskin GS, Fuelberg HE, Holloway JS, Hyer EJ, McMillan WW, Warner J, Streets DG, Zhang Q, Wang Y, Wu S (2010) Source attribution and interannual variability of Arctic pollution in spring constrained by aircraft (ARCTAS, ARCPAC) and satellite (AIRS) observations of carbon monoxide. Atmos Chem Phys 10:977–996. doi:10.5194/acp-10-977-2010

Herber A, Thomason LW, Gernandt H, Leiterer U, Nagel D, Schulz KH, Kaptur J, Albrecht T, Notholt J (2002) Continuous day and night aerosol optical depth observations in the Arctic between 1991 and 1999. J Geophys Res 107(D10):4097. doi:10.1029/2001JD000536

Holben BN, Eck TF, Slutsker I, Tanre D, Buis JP, Setzer A, Vermote E, Reagan JA, Kaufman YJ, Nakajima T, Lavenu F, Jankowiak I, Smirnov A (1998) AERONET—a federated instrument network and data archive for aerosol characterization. Remote Sens Environ 66:1–16

Hudson SR (2011) Estimating the global radiative impact of the sea ice-albedo feedback in the Arctic. J Geophys Res Atmos 116:D16102. doi:10.1029/2011JD015804

Ichoku C, Levy R, Kaufman YJ, Remer LA, Li1 R, Martins VJ, Holben BN, Abuhassan N, Slutsker I, Eck TF, Pietras C (2002) Analysis of the performance characteristics of the five-channel Microtops II sun photometer for measuring aerosol optical depth and precipitable water vapor. J Geophys Res 107:D13. doi:10.1029/2001JD001302

IPCC (2013), climate change 2013, The Physical Science Basis, Stocker TF, Qin D, Plattner GK, Tignor M, Allen SK, Boschung J, Midgley BM (eds.), Contribution of working group I to the fifth assessment report of the intergovernmental panel on climate change.

Labow G (1996) Estimation of ozone with total ozone portable spectroradiometer instruments. II Practical operation and comparisons. Appl Opt 35:6084–6089

Markowicz KM, Flatau PJ, Kardas AE, Remiszewska J, Stelmaszczyk K, Woeste L (2008) Ceilometer retrieval of the boundary layer vertical aerosol extinction structure. J Atmos Oceanic Technol 25(6):928–944

Markowicz KM, Zielinski T, Blindheim S, Gausa M, Jagodnicka AK, Kardas A, Kumala W, Malinowski SP, Petelski T, Posyniak M, Stacewicz T (2012) Study of vertical structure of aerosol optical properties with sun photometers and ceilometer during MACRON Campaign in 2007. Acta Geophys 60(5):1308–1337

Mazzola M, Stone RS, Herber A, Tomasi C, Lupi A, Vitale V, Lanconelli C, Toledano C, Cachorro VE, O'Neill NT, Shiobara M, Aaltonen V, Stebel K, Zielinski T, Petelski T, Ortiz de Galisteo JP, Torres B, Berjon A, Goloub P, Li Z, Blarel L, Abboudm I, Cuevas E, Stock M, Schulz K-H, Virkkula A (2012) Evaluation of sun photometer capabilities for retrievals of aerosol optical depth at high latitudes: The POLAR-AOD intercomparison campaigns. Atmos Environ 52:4–17 (Sp. Iss. SI JUN 2012)

Morys M, Mims III FM, Hagerup S, Anderson SE, Baker A, Kia J, Walkup T (2001) Design, calibration, and performance of Microtops II handheld ozone monitor and sun photometer. J Geophys Res 106:14573–14582

Petelski T, Markuszewski P, Makuch P, Jankowski A, Rozwadowska A (2014) Studies of vertical coarse aerosol fluxes In the boundary layer over the Baltic Sea. Oceanologia 56(4):697–710. doi:10.5697/oc.56-4.697

Petelski T, Piskozub J (2006) Vertical coarse aerosol fluxes in the atmospheric surface layer over the North Polar Waters of the Atlantic. J Geophys Res 111:C06039. doi:10.1029/2005JC003295

Quinn PK, Shaw G, Andrews E, Dutton EG, Ruoho-Airola T, Gong SL (2007) Arctic haze: current trends and knowledge gaps. Tellus B 59(1):99–114. doi:10.1111/j.1600-0889.2006.00238.x

Rozwadowska A, Sobolewski P (2010) Variability in aerosol optical properties at Hornsund, Spitsbergen. Oceanologia 52(4):599–620. doi:10.5697/oc.52-4.599

Rozwadowska A, Górecka I (2012) The impact of a non-uniform land surface on the radiation environment over an Arctic fjord—a study with a 3D radiative transfer model for stratus clouds over the Hornsund fjord, Spitsbergen. Oceanologia 54(4):509–543. doi:10.5697/oc.54-4.509

Rozwadowska A, Zieliński T, Petelski T, Sobolewski P (2010) Cluster analysis of the impact of air back-trajectories on aerosol optical properties at Hornsund, Spitsbergen. Atmos Chem Phys 10 (3):877–893. doi:10.5194/acp-10-877-2010

Shaw GE (1976) Error analysis of multi-wavelength sun photometry. Pure Appl Geophys 114:1–14

Smirnov A, Holben BN, Eck TF, Dubovik O, Slutsker I (2000) Cloud-screening and quality control algorithms for the AERONET database. Remote Sens Environ 73:337–349

Smirnov A, Holben BN, Eck TF, Slutsker I, Chatenet B, Pinker RT (2002) Diurnal variability of aerosol optical depth observed at AERONET (Aerosol Robotic Network) sites. Geophys Res Lett 29(23):2115. doi:10.1029/2002GL016305

Stohl A (2006) Characteristics of atmospheric transport into the Arctic troposphere. J Geophys Res Atmos 111:D11306. doi:10.1029/2005jd006888

Toledano C, Cachorro V, Gausa M, Stebel K, Aaltonen V, Berjon A, Ortis JP, de Frutos AM, Bennouna Y, Blindheim S, Myhre CL, Zibordi G, Wehrli C, Kratzer S, Hakanson B, Carlund T, de Leuww G, Herber A (2012) Overview of sun photometer measurements of aerosol properties in Scandinavia and Svalbard. Atmos Environ 52:18–28. doi:10.1016/j.atmosenv.2011.10.022

Toledano C, Cachorro VE, Berjon A, De Frutos AM, Sorribas M, De la Morena BA, Goloub P (2007) Aerosol optical depth and Ångström exponent climatology at El Arenosillo AERONET site (Huelva, Spain). Quart J Roy Meteor Soc 133(624):795–807

Tomasi C, Vitale V, Lupi A, Di Carmine C, Campanelli M, Herber A, Treffeisen R, Stone RS, Andrews E, Sharma S, Radionov V, von Hoyningen-Huene W, Stebel K, Hansen GH, Myhre CL, Wehrli C, Aaltonen V, Lihavainen H, Virkkula A, Hillamo R, Stroem J, Toledano C, Cachorro VE, Ortiz P, de Frutos AM, Blindheim S, Frioud M, Gausa M, Zielinski T, Petelski T, Yamanouchi T (2007) Aerosols in polar regions: a historical overview based on optical depth and in situ observations. J Geophys Res 112:D16205. doi:10.1029/2007JD008432

Treffeisen R, Herber A, Ström J, Shiobara M, Yamanouchi T, Yamagata S, Holmén K, Kriew M, Schrems O (2011) Interpretation of Arctic aerosol properties using cluster analysis applied to observations in the Svalbard area. Tellus B 56(5). doi:10.3402/tellusb.v56i5.16469

Zawadzka O, Makuch P, Markowicz KM, Zielinski T, Petelski T, Ulevicius V, Strzalkowska A, Rozwadowska A, Gutowska D (2014) Studies of aerosol optical depth with use of Microtops sun photometers and MODIS detectors in the coastal areas of the Baltic Sea. Acta Geophys 62 (2):400–422. doi:10.2478/s11600-013-0182-5

Zielinski T (2004) Studies of aerosol physical properties in coastal areas. Aerosol Sci Technol 38 (5):513–524

Zielinski T, Petelski T, Makuch P, Strzalkowska A, Ponczkowska A, Markowicz KM, Chourdakis G, Georgoussis G, Kratzer S (2012) Studies of aerosols advected to coastal areas with use of remote techniques. Acta Geophys 60(5):1359–1385. doi:10.2478/s11600-011-0075-4

Sea Spray Aerosol Fluxes in the Near Water Boundary Layer—Review of Recent Achievements

Piotr Markuszewski

Abstract This chapter is dedicated to small particles (solid or liquid) generated from the sea surface, so called sea spray aerosol (SSA). Undeniable importance of SSA on climate determines oceanographers and atmospheric scientists to make lots of efforts in resolving all uncertainties connected with measurements, flux parameterizations or transformation processes in the marine boundary layer. A view of the seriousness of the problem gives a fact that the estimated SSA mass emission is still loaded with 80 % relative error. The goal of this text is to briefly introduce the reader to the issue of SSA emission from the sea surface, the main aspects of SSA flux measurements based on selected, most influential articles from recent years.

1 Introduction

The atmospheric aerosol is very important for the geosciences study. Both aerosol impacts: direct effect (absorption and scattering of light jointly called extinction) and its importance as cloud condensation nuclei (indirect effect) is well known and extremely popular in recent literature. Especially interesting and insufficiently explained is the parameterization of aerosol emission from the Earth surface. This study is very important for atmospheric and global climate modeling. The most common aerosol is sea spray aerosol (SSA). 70 % of Earth surface is covered by oceans and seas, that is why the aerosol emission from the sea surface is the highest. According to the newest Intergovernmental Panel on Climate Change (IPCC) report (Boucher et al. 2013) the global emission of SSA is estimated in range 1,400–6,800 (Tg/year), which, as presented in Table 1 dominates all other emissions of the main natural aerosols species, obtained from various model simulations.

P. Markuszewski (✉)
Physical Oceanography Department, Air-Sea Interaction Laboratory, Institute of Oceanology of Polish Academy of Sciences, Powstańców Warszawy 55, Sopot 81-712, Poland
e-mail: pmarkusz@iopan.gda.pl

T. Zielinski et al. (eds.), *Impact of Climate Changes on Marine Environments*,
GeoPlanet: Earth and Planetary Sciences, DOI 10.1007/978-3-319-14283-8_4

Table 1 Global natural emissions of aerosols and aerosol precursors (Boucher et al. 2013)

Source	Natural global	
	Min	Max
Sea spray	1,400	6,800
Including marine POA	2	20
Mineral dust	1,000	4,000
Terrestrial PBAP	50	1,000
Including spores		28
Dimethyl sulphide (DMS)	10	40
Monoterpenes	30	120
Isoprene	410	600
SOA production from all BVOCs	20	380

POA primary organic aerosol, *PBAP* primary biological aerosol particles, *SOA* secondary organic aerosol, *BVOC* biogenic volatile organic compounds

Aside from aerosol origin, it is possible to distinguish classification due to particle sizes. SSA particles exist in two main ranges of sizes, the so called coarse and accumulation modes (particle diameters >2.5 μm, and from 0.1 to 2.5 μm, respectively, Seinfeld and Pandis 2012). The lifetime of SSA in troposphere ranges from 1 day to 1 week. SSA scatters the solar radiation, it has very hygroscopic properties so that means it is very active cloud condensation nuclei (CCN). As regards the chemical composition, according to Seinfeld and Pandis (2012) SSA contains by weight 55.7 % Cl, 0.19 % Br, and 0.00002 % I. The SSA has also primary organic aerosol (POA) compound which is difficult to parameterize (O'Dowd et al. 2004; Gantt et al. 2011; Ovadnevaite et al. 2011; Westervelt et al. 2012; Schmitt-Kopplin et al. 2012). The POA separately has quite different properties. It is observed as Aitken (from 0.01 to 0.1 μm) and accumulation mode and it comes from biologically active marine regions. The POA represents the smallest aerosol particles so they lifetime is approximately 1 week and they are active CCN.

Another very important aerosol species with the second largest global emission is mineral dust. The mineral dust aerosol species exist especially in coarse and super coarse modes. Similarly to SSA mineral dust lives from 1 day to 1 week, depending on particle size and takes part not only in light scattering but also in absorption. It can be active ice nuclei (IN) and takes active role in greenhouse effect (Boucher et al. 2013). Terrestrial primary biological aerosol particles (PBAP) are mainly coarse particles. It has the same lifetime as SSA and mineral dust and may form large CCN and be active IN. Precise description of all atmospheric aerosol compound are presented in (Seinfeld and Pandis 2012).

2 The Main Aspects of the SSA Studies

There are four general areas of SSA study: 1. in situ field measurements, 2. Laboratory Experiments, 3. Remote Sensing, 4. Regional and Global Modeling. Each of these research aspects have their own particular approach and problems. In this part of the article problems connected with each of these issues are shortly presented.

2.1 In Situ Measurements

In situ measurements are probably the most problematic and also the most important aspect of SSA study. The direct aerosol measurements in natural marine environment are the closest to reflecting the real phenomena. Unfortunately, conducting measurements on the open ocean is very expensive and technically difficult to carry. In recent years there were a number of measurement campaigns (inter alia described by: Kulmala et al. 2011; Kleinman et al. 2012; Norris et al. 2012; Lewandowska and Falkowska 2013; Petelski et al. 2014). Kulmala et al. (2011) presented the main achievements of the European Aerosol Cloud Climate and Air Quality Interactions project (EUCAARI) which lasted from 1 January 2007 to 31 December 2010. That big project consisted of multidisciplinary investigations such as observation of physical and chemical composition and optical properties of aerosol particles over Europe, and comprehensive modeling of aerosol processes from nano to global scale and their effects on climate and air quality. Kleinman et al. (2012) presented results from the VOCALS Regional Experiment. In this experiment aircraft observations of aerosol chemistry and physics, stratocumulus clouds properties and atmospheric gaseous composition were conducted. Norris et al. (2012) discussed results of 3 week measurements of marine aerosol fluxes on board the ship in the open Atlantic Ocean region. In this work eddy correlation method was used for wind speed range 4–18 m s^{-1} and size $0.176 < R_{80} < 6.61$ μm. Lewandowska and Falkowska (2013) presented results from observation of SSA chemical composition in the open southern Baltic Sea region and coastal zone. Petelski et al. (2014), presented results of sea spray flux measurements from numerous cruises around the southern Baltic Sea region on board r/v 'Oceania'. Other campaign carried out in the Baltic Sea region is presented by Byčenkienė et al. (2013). Based on the in situ measurements and trajectory-based approach, the Potential Source Contribution Function analysis was performed. Such tool allows to estimate the possible contribution of long-range and local aerosol number concentration transport. All the above mentioned studies are very challenging because of their multidisciplinary character. It is necessary to posses, first of all knowledge in meteorology, atmospheric physic, physical oceanography or biogeochemistry.

There are several aerosol flux measurement methods. De Leeuw et al. (2011), described the following groups of methods:

- Steady State Dry Deposition Method (Smith et al. 1993; Petelski and Piskozub 2006);
- Statistical Wet Deposition Method (Lewis and Schwartz 2004);
- Whitecap Method (Monahan et al. 1986);
- Micrometeorological Methods: eddy correlation (Nilsson and Rannik 2001), gradiental method (Petelski 2003; Petelski and Piskozub 2006; Andreas 2007)
- Multiple Methods (Lewis and Schwartz 2004);

The main disadvantage of all methods, except micrometeorological is the fact that they are based on more qualitative estimation than direct flux determination. However, micrometeorological method and especially eddy correlation (EC) are more and more popular in field measurements over open oceans. Micrometeorological Methods rely on strong physical foundations as Monin-Obukhov theory or Reynolds Decomposition. Eddy covariance method (Lee et al. 2004; Aubinet et al. 2012) is commonly used in air-sea gas transfer measurements, by using high frequency concentration and wind speed measurements ($\sim$20 Hz, commonly used ultrasonic anemometers work even with 50 Hz speed). Unfortunately, it is still impossible to measure aerosol concentration with a frequency as high as in the case of trace gases. However, thanks to technological progress, it is possible to construct increasingly faster particle counters ($\sim$1 Hz is acceptable level, De Leeuw et al. 2007). During SEASAW campaign (described by Norris et al. 2012) there was used A Compact Lightweight Aerosol Spectrometer Probe (CLASP) which allows to measure aerosol concentration with even 10 Hz speed in 0.24 μm $< D_p <$ 18.5 μm range. High frequency measurements are necessary to record air turbulence spectrum. In fluid dynamics there is common mathematical technique, which allow to separate average $\bar{x}$ and fluctuating x' parts of given quantity (Müller 2006; Foken and Nappo 2008):

$$x = \bar{x} + x' \tag{1}$$

In turbulent flow with assumptions of negligible density fluctuations, negligible mean vertical flow (no divergence/convergence), there is possible to determine the net flux of each meteorological parameter as:

$$F \approx \overline{\rho_a w' s'} \tag{2}$$

where ρ_a is an air density, $\overline{w's'}$ is a covariance of vertical fluctuating component of wind (w′) and fluctuation of a given meteorological parameter s′.

The Gradient method (GM), in contrast to the EC, relies on continuous measurements of aerosol concentration on at least three levels. To calculate aerosol flux based on the M-O theory, there is an assumption of the particle concentration as a scalar property of the air. Based on this and under the condition of horizontal uniformity, vertical flux equals to the emission from the sea surface. It is possible to fully determine horizontal uniformity by using such parameters as momentum flux τ, sensible heat flux Q and buoyancy parameter $\beta = g/T$ (g-gravitational acceleration, T-air temperature). These parameters allow to define following scales: friction velocity:

$u_* = (\tau/\rho)^{1/2}$, Temperature: $T_* = -Q/\kappa u_*$ and Length: $L = -u^3/\kappa\beta Q$ (ρ-density of the air, κ-Von Kármán constant. The scale of particle concentration is defined as:

$$N_* = F_N/u_* \tag{3}$$

where: F_N is the aerosol flux.

It is possible to express the non-dimensional aerosol concentration gradient by the universal function of z/L:

$$(z/N_*)\partial \mathrm{N}/\partial \mathrm{z} = \Phi(z/L) \tag{4}$$

Using Eq. 4 it is possible to derive the final equation using asymptotic forms from the M-O theory. The most popular for near water atmospheric boundary layer is the following formula:

$$N(z) = N_* \ln(\mathrm{z}) + C \tag{5}$$

Measurements of concentration on 3 levels above sea surface, allow to calculate N_* and thus aerosol fluxes. In measurements presented by Petelski (2003), Petelski and Piskozub (2006), Petelski et al. (2014), there are presented GM measurements on board s/y Oceania, where there is Classical Aerosol Spectrometer (CSASP-100-HV, Zielinski 2004) used. The probe is placed on a special lift on board of the vessel. The aerosol concentration is measured on five levels above sea surface 8, 11 14, 17 and 20 m. Another newest achievement using gradient method is presented in Savelyev et al., (2014). In this paper there are successfully compared in situ measurements of aerosol production (GM and dry deposition) and direct passive microwave remote sensing.

2.2 Laboratory Experiments

The aim of laboratory experiments is to develop knowledge of SSA emission processes. SSA is generated from the sea surface as water drops through several processes. The collapsing wind waves are the main mechanism in which the SSA is transported to the atmosphere. Therefore, the emission depends on amount of wind wave energy, dissipated in the breaking process. Such phenomenon is however, very difficult to parameterize (Massel 2007).

The nature of aerosol emission is strongly correlated with wind speed. For wind speed in range from 5 to 10 m/s emission from bursting bubbles created during wave collapsing (so-called film and jet drops) is the dominating process. In higher wind speed conditions, the spume tearing from wave crests (spume drops) dominates the emission. This process generates the largest aerosol droplets (reaching even the 1,000 μm in radius). The secondary process consists of large droplets falling to the sea surface and creating smaller drops within the impact (splash drops).

Laboratory experiments allow scientists to study all mechanisms in an isolated system. In this approach this is possible to better understand each single mechanism. With such unquestionable advantage, there are still some issues bothering the scientists, which include problems with standardization and representativeness the measurements (for example standardization of sea water, Meskhidze et al. 2013). There are also problems with terminology unification.

With all advantages and disadvantages laboratory experiments are very important in the SSA flux study. Through such investigation it is possible to find the most universal parameter describing the SSA emission.

The most influential articles in this subject from recent years concentrate mainly on contribution of organic compounds to SSA chemical, physical and size distribution properties. Fuentes et al. (2010, 2011) presented results of investigating impacts of phytoplankton on properties of primary marine aerosol and source fluxes. Park et al. (2014a) investigate marine aerosol production based on measurements of insoluble submicrometer particles and biological materials in sea water. In this article influence of anthropogenic contribution on size distribution is described along with observed seasonal variability of marine aerosol concentration changes with biological composition of sea water. Park et al. (2014b) investigated mixing state of submicrometer SSA and measured differences between natural sea water with artificial water in the laboratory environment. Prather et al. (2013) presented results of newly implemented approach of laboratory experiment so called mesocosm experiment. In this experiment it is possible to reproduce the chemical complexity of SSA in artificial environment. For natural seawater, SSA concentration variability was measured, depending on parameters such as phytoplankton and heterotrophic bacteria concentration or chlorophyll-*a*.

2.3 Remote Sensing

Remote sensing consists of all passive measurements, based on extinction (absorption and scattering) of electromagnetic radiation by aerosol particles, in given area or cross section. The most important remote sensing systems investigating the aerosol include:

- Maritime Aerosol Network (Smirnov et al. 2009), based on aerosol optical depth (AOD) measurements using Microtops II sun photometers on board the ships over the World's Oceans.
- Aerosol Robotic Network, is the international program involving many ground remote sensing aerosol networks established by NASA and PHOTONS (PHOtométrie pour le Traitement Opérationnel de Normalisation Satellitaire). This federation gathers institutes, national agencies, individual scientist all over the world. Collaboration via AERONET provides observation of AOD and all related meteorological or other aerosol properties parameters.

- Moderate resolution Imaging Spectroradiometer (MODIS), one of the most important remote sensing device. Involves two satellites: Terra and Aqua. This system is designed for large-scale global dynamics monitoring. Placed into the orbit by NASA.
- Multi-angle Imaging Spectral Radiometer (MISR), another satellite in this measurements unique is that the instrument is equipped in cameras pointed in 9 different directions. As a result, it is possible to gain very accurate information about radiation coming from earth. Also launched by NASA.
- Advanced Along Track Scanning Radiometer (AATSR), main purpose of this satellite is the global Sea Surface Temperature (SST) monitoring. Besides SST this system provides data about surface temperature in general, clouds, aerosols, vegetation and snow. Belongs to European Space Agency (ESA).
- Polarization and Directionality of the Earth's Reflectances (POLDER), another satellite developed by French space agency CNES.
- Medium-spectral Resolution Imaging Spectrometer (MERIS), device placed on board Envisat platform put into the orbit by ESA.
- Sea-viewing Wide Field-of-view Sensor (SeaWiFS), project based on the satellite observations designed to collect global ocean surface biological data. Another system provide by NASA.
- Cloud-Aerosol Lidar and Infrared Pathfinder Observations (CALIPSO), involves lidar measurements and passive infrared and visible observations from satellites to investigate vertical profile of atmosphere.

Based on comprehensive (ground, satellite, aircraft-based) measurements such interdisciplinary kind of approach, allows to estimate not only the global AOD distribution or the SSA emission, but also budget of SSA, surface wind speed, surface wave phase velocity, Chl-*a* concentration, colored dissolved organic matter (CDOM) concentration whitecap fraction and sea surface temperature (SST), (Meskhidze et al. 2013). The reason for such wide-ranging measurements is to estimate and compare the influence of above parameters on aerosol emission from the marine surface. Ground remote measurements are necessary to validate satellite measurements (Melin et al. 2013).

2.4 Regional and Global Modeling

More and more models of climate or atmosphere have implementation of interactive sea-salt aerosol emission part. The role of such modeling is to estimate the global emission of SSA to the atmosphere. Very important is also prediction of SSA influence on weather and cloud-creative processes. The main problem for such research based on modeling is lack of in situ data. It is very important to compare theoretical predictions with real measurements. The most popular data from direct measurements is the mass concentration measurements. For improvement quality of modeling processes connected with SSA there are plans to develop large comprehensive databases with in situ measurement data (Meskhidze et al. 2013). Example of such data base is the data base of the University of Miami (Mishchenko et al. 2002). It is essential to gather long data base, to compare modeled data with observations of the seasonal variability and trends.

Other subject is a choice of parameters used in models. The main and easiest to apply is the wind speed at 10 m above sea surface. It is the simplest meteorological parameter to use, because of direct relation between surface wave height or whitecap generation. Science community still discuses about using other parameters instead. The main other parameters are (Lewis and Schwartz 2004): Atmospheric Stability, Wind Friction Velocity, Sea Water Temperature, Wave Phase Velocity, Fetch, Salinity, Surface-Active Substances. Atmospheric Stability strongly affects air mass movements so there is strong dependence for marine aerosol drops coming from whitecaps. Wind friction velocity influences not only stability of atmosphere but also whitecap ratio. This parameter is quite easy to obtain, so it is convenient using it during source function estimation or model execution. Sea water temperature is strongly related to the kinetic viscosity of seawater so also existing of whitecaps also should be related. The state of sea surface has its specific inertia. Parameter which informs about state of the sea surface is wave phase velocity. Other parameter connected with sea surface inertia is fetch. Fetch is the distance over water that the wind has blown and may affect the wind spectrum. There is no direct relation between aerosol emission and salinity. However, there are differences in observations for brackish seas and open oceans with fresh water. Surface active substances may change the sea state, roughness length, surface tension of the seawater-air layer and affect on the lifetime of foam on the sea surface. Unfortunately, all these relations are very difficult to parameterize. The very interesting parameterization was presented by Ovadnevaite et al. (2014) where the Reynolds Number was used instead of all the above parameters. Advantage of such approach is that Reynolds Number brings information about wind speed, kinematic viscosity of water and indirectly: wave height, wind history, friction velocity or viscosity.

The new approach for determining SSA fluxes is presented by Grythe et al. (2013). In this paper I have reviewed 21 SSA source functions known from the literature. For each function a global SSA emission was described. In applying this task the FLEXPART Lagrangian particle dispersion model was used. Additionally, the authors proposed a new source function. This function, based on modeling estimation determined the functional relation of the SSA emission versus such parameters as wind speed (power dependence, $\sim u^{3.5}$) or sea surface temperature and aerosol diameter ($D_p < 10$ μm, lognormal relation enclosing 3 aerosol modes). This is the first function determined using modeling estimation. The comparison of the SSA emission obtained from all 21 source functions is presented in Table 2 (Grythe et al. 2013).

Table 2 Comparison of the source functions used by Grythe et al. (2013) with selected components

Reference	D_p	Type	Pg year^{-1}	±
Monahan et al. (1986)	0.8–8	Exp.	4.51	0.44
Sofiev et al. (2011)	0.01–10	Modified	5.87	0.57
Sofiev et al. (2011)	0.01–10	Modified	1.83	0.18
Gong (2003)	0.07–20	Lab.	5.95	0.58
Clarke et al. (2006)	0.01–8	Surf exp.	22.6	2.19
Sofiev et al. (2011)	0.01–10	Modified	2.59	0.33
Gong (2003)	0.07–20	Modified	4.59	0.57
Monahan et al. (1986)	0.1–10	Exp.	5.20	0.50
Jaeglé et al. (2011)	0.07–20	Model	4.86	0.34
Jaeglé et al. (2011)	0.07–20	Model	4.20	0.39
	<20	Model	17.43	1.01
Smith et al. (1993)	0.3–25	Exp.	2.90	0.20
Smith and Harrison (1998)	1–300	Dry dep.	6.67	0.66
Lewis and Shwartz (2004)	1–25	Multiple	73.53	5.82
Andreas (1998)	1–20	Modified	10.14	0.69
Andreas (1990)	0.08–15	Multiple	605	43.8
De Leeuw et al. (2000)	0.8–10	Surf Exp.	2,444	491
Andreas (1992)	0.08–15	Modified	5.65	0.45
Petelski and Piskozub (2006)	0.25–7.5	Exp.	167.8	0.92
Andreas (2007)	0.25–7.5	Modified	7.09	14.44
Norris et al. (2008)	<2.4	Field.	3.25	0.68
Grythe et al. (2013)	0.01–10	Model	8.91	0.61

D_p is a dry diameter. "Type" is the source function determination method. The fourth column presents the annual averaged global production from over 25 years. The data comes from the European Centre for Medium-Range Weather Forecasts (ECMWF). The last column presents difference in mass between maximum and minimum year, given in Pg

3 Summary

In this chapter I have described the latest efforts in sea spray aerosol flux studies. The SSA fluxes have a great influence on air-sea interaction processes and thus the climate processes. Several aspects of the studies on the SSA transfer have still low scientific understanding. Meskhidze et al. (2013) presented an understanding level of each SSA study aspects (Table 3).

The main components with the lowest understanding and the greatest impact, according to Meskhidze et al. (2013) include:

1. Size-resolved chemical composition/hygroscopicity,
2. Mixing state
3. Wet removal,
4. Photochemical aging,

Table 3 Prioritization matrix by Meskhidze et al. (2013)

Parameter	Current understanding	Impact if achieved	Difficulty/ resources needed
Source function			
Bulk mass and number emissions	Med	Med-Low	Med-Low
Bulk chemical composition/ hygroscopicity	Med-Low	Med-Low	Med-Low
Size-resolved mass and number emissions	Med-Low	High	Med
Size-resolved chemical composition/ hygroscopicity	Low	High	High
Mixing state	Low	High	High
CCN number flux	Med-Low	High	Med
Giant CCN	Low	Med	High
IN number emissions	Low	Med	Med
IN sources	Low	Med	Med
Whitecap fraction	Med	High	Med-Low
Bubble spectra	Med	High	Med
Seawater/microlayer chemical composition	Low	High	Med
Size-resolved organic speciation	Low	High	High
Optical properties			
AOD	Med	High	Med-Low
Refractive index	Med	Med	Med-Low
Depolarization	Med	Med-Low	High
Humidified scattering	Med	High	Low
Ångström exponent	Med	Med-Low	Med
Absorption Ångström exponent	Med-Low	Med-Low	Med
Fluorescence	Med-Low	Med	Med-Low
Lidar ratio	Med-Low	Med	Med
Marine boundary layer budget			
Wet removal	Low	High	Med
Dry removal	Med-Low	Med	Med
Photochemical aging	Low	High	High
Volatility	Low	Med	Med
Entrainment	Med-Low	High	Med
Cloud processing	Low	High	High
Transport	Med-Low	Med	Med

Ranking levels were assigned using the following numerical values: 1: Low, 2: Med-Low, 3: Med, 4: Med-High, and 5: High

5. Cloud processing,
6. Seawater/microlayer chemical composition,
7. Size resolved organic speciation.

Only by conducting interdisciplinary measurements and research, it will be possible to provide a comprehensive description of marine boundary layer physics.

Acknowledgments This research has been partly made within the framework of a Polish-Norwegian Research Programme operated by the National Centre for Research and Development under the Norwegian Financial Mechanism 2009–2014 in the frame of Project Contract No Pol-Nor/196911/38/2013, partly within the scope of the GAME project.

References

Andreas EL (1990) Time constants for the evolution of sea spray droplets. Tellus B 42:481–497
Andreas EL (1992) Sea spray and the turbulent air–sea heat fluxes. J Geophys Res 97:11429–11441
Andreas EL (1998) A new sea spray generation function for wind speeds up to 32 ms^{-1}. J Phys Oceanogr 28:2175–2184
Andreas EL (2007) Comments on "Vertical coarse aerosol fluxes in the atmospheric surface layer over the North Polar Waters of the Atlantic" by Tomasz Petelski and Jacek Piskozub. J Geophys Res 112:C11010. doi:10.1029/2007JC004184
Aubinet M, Vesala T, Papale D (2012) Eddy covariance a practical guide to measurement and data analysis. Springer Atmospheric Sciences, Germany
Boucher O, Randall D, Artaxo P, Bretherton C, Feingold G, Forster P, Kerminen V-M, Kondo Y, Liao H, Lohmann U, Rasch P, Satheesh SK, Sherwood S, Stevens B, Zhang XY (2013) Clouds and aerosols. In: Stocker TF, Qin D, Plattner G, Tignor M, Allen SK, Boschung J, Nauels A, Xia Y, Bex V, Midgley PM (eds) Climate change 2013: the physical science basis. Contribution of working group i to the fifth assessment report of the intergovernmental panel on climate change. Cambridge University Press, Cambridge
Byčenkienė S, Ulevicius V, Prokopčiuk N, Jasinevičienė D (2013) Observations of the aerosol particle number concentration in the marine boundary layer over the south-eastern Baltic Sea. Oceanologia 55(3):573–598. doi:10.5697/oc.55-3.573
Clarke AD, Owens SR, Zhou J (2006) An ultrafine sea salt flux from breaking waves: implications for cloud condensation nuclei in the remote marine atmosphere. J Geophys Res 111:D06202. doi:10.1029/2005JD006565
De Leeuw G, Neele FP, Hill M, Smith MH, Vignati E (2000) Sea spray aerosol production by waves breaking in the surf zone. J Geophys Res 105:29397–29409
De Leeuw G, Moerman M, Zappa CJ, McGillis WR, Norris S, Smith M (2007) Eddy correlation measurements of sea spray aerosol fluxes. Transport at the air-sea interface. Springer, Berlin, Heidelberg, pp 297–311
De Leeuw G, Andreas EL, Anguelova MD, Fairall CW, Lewis ER, Dowd CO, Schulz M, Schwartz SE (2011) Production flux of sea spray aerosol. Rev Geophys 49:1–39
Foken T, Nappo CJ (2008) Micrometeorology. Springer, New York
Fuentes E et al (2010) On the impacts of phytoplankton-derived organic matter on the properties of the primary marine aerosol–Part 1: Source fluxes. Atmos Chem Phys 10(19):9295–9317
Fuentes E, Coe H, Green D, McFiggans G (2011) On the impacts of phytoplankton-derived organic matter on the properties of the primary marine aerosol–Part 2: Composition, hygroscopicity and cloud condensation activity. Atmos Chem Phys 11(6):2585–2602

Gantt B, Meskhidze N, Facchini MC, Rinaldi M, Ceburnis D, O'Dowd CD (2011) Wind speed dependent size-resolved parameterization for the organic mass fraction of sea spray aerosol. Atmos Chem Phys 11(16):8777–8790

Gong SL (2003) A parameterization of sea salt aerosol source function for sub and super micron particles. Global Biogeochem Cycles 17:1097. doi:10.1029/2003GB002079

Grythe H, Ström J, Krejci R, Quinn P, Stohl A (2013) A review of sea-spray aerosol source functions using a large global set of sea salt aerosol concentration measurements. Atmos Chem Phys 14:1277–1297. doi:10.5194/acp-14-1277-2014

Jaeglé L, Quinn PK, Bates TS, Alexander B, Lin J-T (2011) Global distribution of sea salt aerosols: new constraints from in situ and remote sensing observations. Atmos Chem Phys 11:3137–3157. doi:10.5194/acp-11-3137-2011

Kleinman LI, Daum PH, Lee YN, Lewis ER, Sedlacek AJ III, Senum GI, Allen G (2012) Aerosol concentration and size distribution measured below, in, and above cloud from the DOE G-1 during VOCALS-REx. Atmos Chem Phys 12(1):207–223

Kulmala M, Asmi A, Lappalainen HK, Baltensperger U, Pandis SN (2011) General overview: European integrated project on aerosol cloud climate and air quality interactions (EUCAARI) —integrating aerosol research from nano to global scales. Atmos Chem Phys 11:13061–13143. doi:10.5194/acp-11-13061-2011

Lee X, Massman W, Law B (2004) Handbook of micrometeorology. Kluwer Academic Publishers, Dordrecht

Lewandowska AU, Falkowska LM (2013) Sea salt in aerosols over the southern Baltic. Part 1. The generation and transportation of marine particles. Oceanologia 55(2):279–298. doi:10.5697/oc.55-2.279

Lewis ER, Schwartz SE (2004) Sea salt aerosol production: mechanisms, methods, measurements and models—a critical review. Geophys Monogr Ser 152:413. AGU, Washington, D. C

Massel SR (2007) Chapter 9 Marine aerosol fluxes, in ocean waves breaking and marine aerosol fluxes. Springer, New York, pp 229–246

Melin F, Zibordi G, Carlund T, Holben BN, Stefan S (2013) Validation of SeaWiFS and MODIS Aqua-Terra aerosol products in coastal regions of European marginal seas. Oceanologia 55(1). doi:10.5697/oc.55-1.027

Meskhidze N, Petters MD, Tsigaridis K, Bates T, O'Dowd C, Reid J, Zorn SR (2013) Production mechanisms, number concentration, size distribution, chemical composition, and optical properties of sea spray aerosols. Atmos Sci Lett 14(4):207–213

Mishchenko M, Penner J, Anderson D (2002) Global aerosol climatology project. J Atmos Sci 59:249

Monahan EC, Spiel DE, Davidson KL (1986) A model of marine aerosol generation via whitecaps and wave disruption. In: Monahan EC, MacNiocaill G (eds) Oceanic whitecaps and their role in air-sea exchange processes. Reidel, Dordrecht, pp 167–174

Müller P (2006) The equations of oceanic motions. Cambridge University Press, Cambridge

Nilsson ED, Rannik Ü (2001) Turbulent aerosol fluxes over the Arctic Ocean: 1. Dry deposition over sea and pack ice. J Geophys Res 106:32125–32137. doi:10.1029/2000JD900605

Norris SJ, Brooks IM, de Leeuw G, Smith MH, Moerman M, Lingard JJN (2008) Eddy covariance measurements of sea spray particles over the Atlantic Ocean. Atmos Chem Phys 8:555–563. doi:10.5194/acp-8-555-2008

Norris SJ, Brooks IM, Hill MK, Brooks BJ, Smith MH, Sproson DAJ (2012) Eddy covariance measurements of the sea spray aerosol flux over the open ocean. J Geophys Res Atmos 117: D07210. doi:10.1029/2011jd016549

O'Dowd CD, Facchini MC, Cavalli F, Ceburnis D, Mircea M, Decesari S, Putaud JP (2004) Biogenically driven organic contribution to marine aerosol. Nature 431(7009):676–680

Ovadnevaite J, O'Dowd C, Dall'Osto M, Ceburnis D, Worsnop DR, Berresheim H (2011) Detecting high contributions of primary organic matter to marine aerosol: a case study. Geophys Res Lett 38(2):L02807

Ovadnevaite J, Manders A, De Leeuw G, Ceburnis D, Monahan C, Partanen AI, O'Dowd CD (2014) A sea spray aerosol flux parameterization encapsulating wave state. Atmos Chem Phys 14(4):1837–1852

Park JY, Kim M, Han S, Lim S, Kim G, Park K (2014a) Measurement of insoluble submicrometer particles and biological materials in seawater to investigate marine aerosol production. J Aerosol Sci 75:22–34

Park JY, Lim S, Park K (2014b) Mixing state of submicrometer sea spray particles enriched by insoluble species in bubble-bursting experiments. J Atmos Ocean Technol 31(1):93–104

Petelski T (2003) Marine aerosol fluxes over open sea calculated from vertical concentration gradients. J Aerosol Sci 34:359–371

Petelski T, Piskozub J (2006) Vertical coarse aerosol fluxes in the atmospheric surface layer over the North Polar Waters of the Atlantic. J Geophys Res 111:C06039. doi:10.1029/2005JC003295

Petelski T, Markuszewski P, Makuch P, Jankowski A, Rozwadowska A (2014) Studies of vertical coarse aerosol fluxes in the boundary layer over the Baltic Sea. Oceanologia 56(4). doi:10.5697/oc.56-3.651

Prather KA, Bertram TH, Grassian VH, Deane GB, Stokes MD, DeMott PJ, Zhao D (2013) Bringing the ocean into the laboratory to probe the chemical complexity of sea spray aerosol. Proc Natl Acad Sci 110(19):7550–7555

Savelyev IB, Anguelova MD, Frick GM, Dowgiallo DJ, Hwang PA, Caffrey PF, Bobak JP (2014) On direct passive microwave remote sensing of sea spray aerosol production. Atmos Chem Phys 14(21): 11611–11631

Schmitt-Kopplin P, Liger-Belair G, Koch BP, Flerus R, Kattner G, Harir M, Herndl G (2012) Dissolved organic matter in sea spray: a transfer study from marine surface water to aerosols. Biogeosciences 9(4):1935–1955

Seinfeld JH, Pandis SN (2012) Atmospheric chemistry and physics: from air pollution to climate change. Wiley, New York

Smirnov A, Holben BN, Slutsker I, Giles DM, McClain CR, Eck TF, Jourdin F (2009) Maritime aerosol network as a component of aerosol robotic network. J Geophys Res Atmos (1984–2012) 114(D6)

Smith MH, Harrison NM (1998) The sea spray generation function. J Aerosol Sci 29:189–190. doi:10.1016/S0021-8502(98)00280-8

Smith MH, Park PM, Consterdine IE (1993) Marine aerosol concentrations and estimated fluxes over the sea. Q J R Meteorol Soc 119(512):809–824

Sofiev M, Soares J, Prank M, de Leeuw G, Kukkonen J (2011) A regional-to-global model of emission and transport of sea salt particles in the atmosphere. J Geophys Res 116:D21302. doi:10.1029/2010JD014713

Westervelt DM, Moore RH, Nenes A, Adams PJ (2012) Effect of primary organic sea spray emissions on cloud condensation nuclei concentrations. Atmos Chem Phys 12(1):89–101

Zielinski T (2004) Studies of aerosol physical properties in coastal areas. Aerosol Sci Technol 38 (5):513–524

Acoustical and Optical Methods in Arctic Zooplankton Studies

Lukasz Hoppe and Joanna Szczucka

Abstract Concurrent acoustical and optical measurements have a great potential to describe zooplankton distribution over large temporal and spatial scales. It is difficult to collect complete information on zooplankton distribution with traditional methods (e.g. nets), that provide discrete and low resolution data on distribution of zooplankton biomass, abundance, as well as community structure of zooplankton. Acoustic sounding makes environmental studies fast, non-intrusive, and relatively cheap with high temporal and spatial resolution. LOPC delivers real-time information on zooplankton abundance and size spectra. In this review we present the results of study on zooplankton distribution in two fjords of Spitsbergen in the summer of 2013. Data for this study was collected during simultaneous profiling with high frequency (420 kHz) echosounder and LOPC along the main fjord axes. Zooplankton size spectra obtained by LOPC were used as input parameters in "high-pass" model of sound scattering on fluid-like particles. Model output values of acoustic backscattering strength were compared with values obtained by echosounding. In most cases there was a good agreement between measured and modeled values, except conditions of very low zooplankton abundance and events of fish presence. Zooplankton size structure is helpful in validating and refinement of "high-pass" acoustic model for specific set of scatterers. This gives a possibility to determine the theoretical backscattering strength of zooplankton. Implementing two complementary methods allows to obtain fast and more complete information on zooplankton distribution.

Keywords High-frequency acoustics · Arctic zooplankton · LOPC · Sound scattering model

L. Hoppe (✉) · J. Szczucka
Institute of Oceanology, Polish Academy of Sciences, ul. Powstańców Warszawy 55, 81-712 Sopot, Poland
e-mail: hoppe@iopan.gda.pl

T. Zielinski et al. (eds.), *Impact of Climate Changes on Marine Environments*, GeoPlanet: Earth and Planetary Sciences, DOI 10.1007/978-3-319-14283-8_5

1 Introduction

This research focused on zooplankton, a key component of the ecosystem linking primary producers with higher trophic levels. Zooplankton community is characterized by irregular distribution with patches caused by either environmental or behavioral factors (Omori and Hamner 1982). Because it is typically difficult to assess detailed information on zooplankton distribution with traditional methods (e.g. nets), use of alternative methods should be taken into consideration. Acoustic sampling makes studies of ecosystems fast, non-intrusive, and relatively cheap, with high spatial and temporal resolution (Wade and Heywood 2001). Optical method makes it possible to assess zooplankton distribution, abundance and community size spectra, and has been already proved in practical applications (Trudnowska et al. 2012; Krupica et al. 2012).

Concurrently with high frequency acoustical measurements, complementary methods were used—Laser Optical Plankton Counter (LOPC) measurements delivered high resolution information on zooplankton size spectra and abundance, together with environmental parameters.

Information on population size spectra allowed implementation of the mathematical model of acoustic scattering on the Arctic zooplankton community. The last part of research was a comparison between measured values of backscattered acoustic energy and model-calculated results.

The results presented below were obtained during research conducted in two fjords of Spitsbergen in the summer of 2013.

2 Study Area

Zooplankton acoustical and optical studies were carried out in two West Spitsbergen fjords, Hornsund and Kongsfjorden, during cruise of r/v Oceania in the summer of 2013, as a part of the GAME (Growing of the Arctic Marine Ecosystem) project. Hornsund is regarded as a cold fjord, under the influence of the South Cape Current, while Kongsfjorden is the fjord influenced by warmer Atlantic type waters, originating from the West Spitsbergen Current (Fig. 1).

3 Materials and Methods

Acoustic measurements were done with a DT-X echosounder (BioSonics Inc., Seattle, USA), working at a frequency of 420 kHz. Downward-looking acoustic transducer was mounted on a frame attached to the ship, such that the transducer was submerged 1 m below the surface. The construction allowed the ship to sail with a speed of approximately 3 knots. Pulse length was set at 0.3 ms and trigger

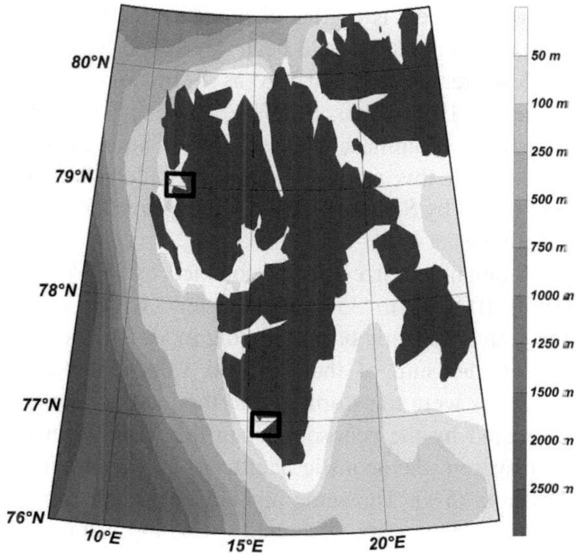

Fig. 1 Area of investigation —West Spitsbergen fjords: Kongsfjorden (north) and Hornsund (south)

value was fixed at 2 Hz. The output of the echosounder was a volume backscattering strength S_V, which is a logarithmic measure of the volume backscattering coefficient s_V, being the sum of the backscattering cross-sections of all scatterers enclosed in the ensonified water volume. The output S_V values were averaged over 1 m depth layers and 5 s time intervals (including 10 transmissions). The 420 kHz transducer produces an acoustic signal of about 3 mm wave length, thus it is possible to detect with this instrument individual zooplankters with the equivalent spherical radius of 0.5 mm, assuming the criterion of detectability: 2π radius/wavelength > 1 (Medwin and Clay 1998).

Concurrently with the echosounding, Laser Optical Plankton Counter—LOPC (Brook Ocean Technology Dartmouth, Canada) was hauled alongside the ship, performing continuous measurements during cruising. LOPC is the in situ instrument which provides data on abundance and size structure of particles and plankton in the water. The system used in this study allows us to obtain information on size spectra for particles (including particulate matter and plankton) in the size range from 100 μm to 35 mm Equivalent Spherical Diameter (ESD). LOPC was working in an undulating mode from the surface to 25 m depth (occasionally reaching 50 m depth) with the pulse rate fixed at 2 Hz. Additionally, during the LOPC survey, distribution of other environmental parameters such as salinity, temperature and fluorescence, were measured by the probes mounted on the same towing body.

Establishing quantitative relationship between the acoustic backscattered signal and the zooplankton abundance, its size and taxonomic structure is a complex problem. The acoustic scattering intensity depends on the wave frequency and the scatterers' size, shape, age, concentration, orientation, and material properties (sound speed and density ratios). Because of this complexity, development of acoustic models is a great challenge. Many different models, which vary in accuracy and generality, have been developed within the past decades. First scattering models treated zooplankton as a homogeneous fluid sphere (Rayleigh 1945; Anderson 1950; Johnson 1977). More sophisticated models take shape and material properties of animals (Stanton 1989; Chu et al. 1992; Stanton and Chu 2000) into account. In this study we used so called "high-pass" model for a fluid sphere introduced by Stanton (1989), which proved to work well in the Arctic environment (Trudnowska et al. 2012).

Zooplankton size spectra obtained by LOPC were divided into 49 size classes (from 100 μm to 35 mm diameter), which were used as input parameters for theoretical acoustic backscattering model of zooplankton. To validate implemented model, the sum of the modeled values of backscattering cross-sections of all scatterers were transformed into the total S_V and compared with the real S_V values measured by the echosounder in the water layer of the LOPC towing depth. For comparison, the acoustic data were averaged over 1 m depth layers and 10 transmissions (5 s). The same applied to the optical data and the environmental variables, which were averaged over the same time and depth intervals. The zooplankton sound speed contrast was set at $h = 1.027$ and the density contrast at $g = 1.0$, as proposed for the Arctic copepod species *Calanus finmarchicus* (Kogeler et al. 1987). The collected data consist of, in total, 13 h of concurrent acoustical and optical measurements (5 h in Hornsund and 8 h in Kongsfjorden), which were organized into 30 min long data series.

4 Results

Two exemplary data series were selected for this report, to illustrate the results of our study. The first is the data collected in Kongsfjorden on 06 August 2013, starting at 18:49 UTC. The distribution of backscattering strength in the 0–50 m layer during the 30 min long survey along the designated line is shown in Fig. 2, with the LOPC towing route indicated. Distribution of backscatter presumably reflects the distribution of biomass, most likely of zooplankton, with incidental strong echoes from other objects like fish. For each time interval the backscattering strength measured by the 420 kHz echosounder (red curve) was compared with the backscattering strength calculated from the model (blue curve), for the zooplankton size and abundance estimations based on the optical measurements at the exact towing depth (Fig. 3). The repetitive maxima and minima reflect changes in S_V values, caused by higher or lower zooplankton biomass at different water depths. The linear correlation coefficient between the measured and modeled values of S_V for the whole 30 min long data set in the entire studied water column equaled to $R = 0.72$ (Fig. 4).

The data series taken in Hornsund on 26 July 2013, starting at 21:26 UTC illustrates the technical capability of measuring several marine environment properties simultaneously with optical data. Data on distribution of various

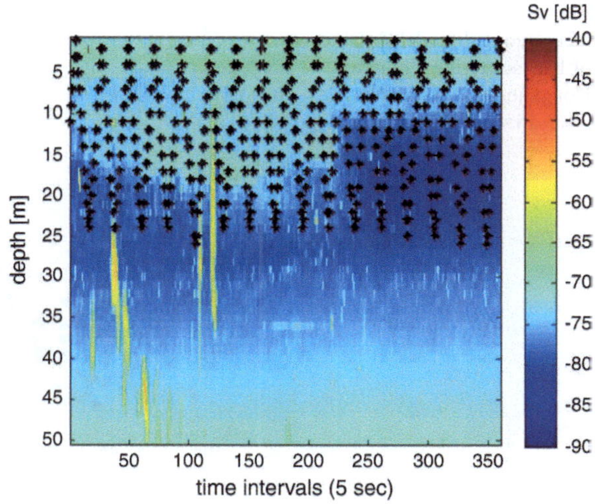

Fig. 2 Echogram with LOPC towing route during the undulating mode (black asterisks); Kongsfjorden, 06 August 2013 18:49 UTC

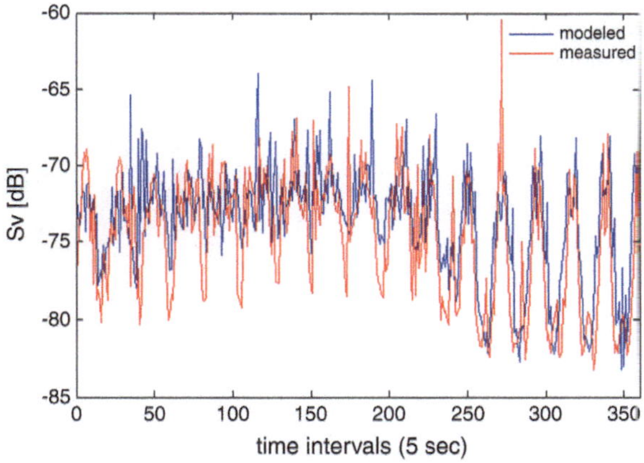

Fig. 3 A comparison of the measured and modeled values of backscattering strength averaged over 1 m depth layer and 10 transmissions (5 s); Kongsfjorden, 06 August 2013 18:49 UTC

environmental factors increases the possibility to interpret the study results and to understand the complexity of ecosystem. This series includes the data on distribution of acoustic backscattering, particles from the LOPC surveys, as well as data on distribution of temperature, salinity and fluorescence, which were collected together with the LOPC data. The distribution of temperature (Fig. 5) clearly shows the presence of hydrological front located in the middle of survey section. The LOPC data, at the same time, pronounced differences in distribution of particles, with high abundance of suspended matter in the colder waters which presumably originated from glacier (Fig. 6). Elevated levels of acoustic backscatter were also recorded in the same fragment of the study section (Fig. 7). A comparison of measured and modeled values of backscattering strength indicates that in the glacial waters there is lack of agreement between two methods (Fig. 8). Modeled values are much lower than the measured ones.

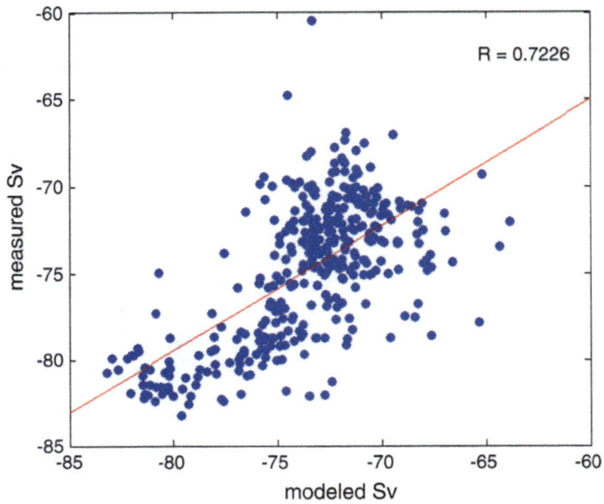

Fig. 4 Correlation of the measured and modeled values of backscattering strength; Kongsfjorden, 06 August 2013 18:49 UTC

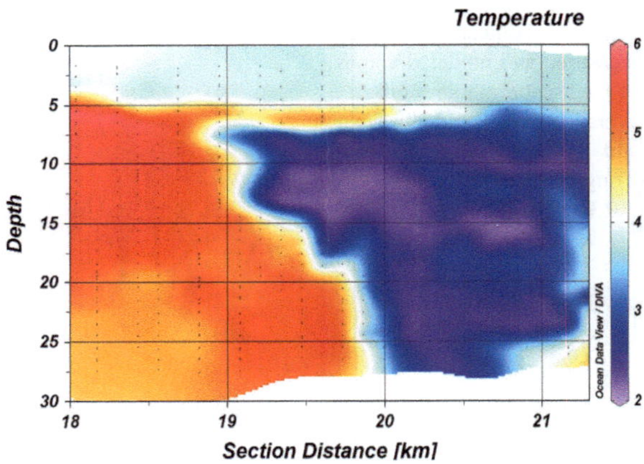

Fig. 5 Temperature [°C] plot for a 30 min long transect; Hornsund, 26 July 2013 21:26 UTC

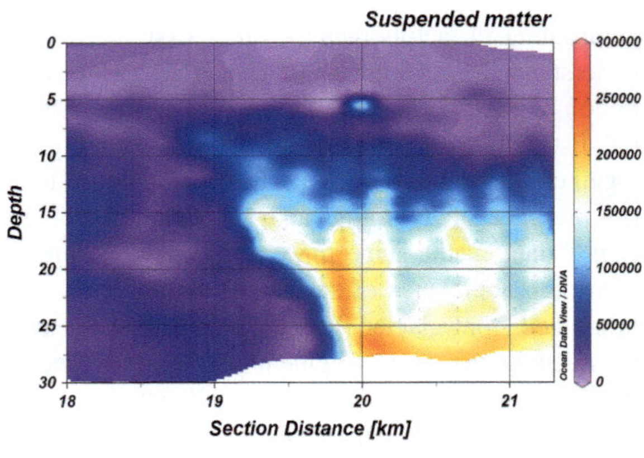

Fig. 6 Suspended matter [ind/m^3] plot for a 30 min long transect; Hornsund, 26 July 2013 21:26 UTC

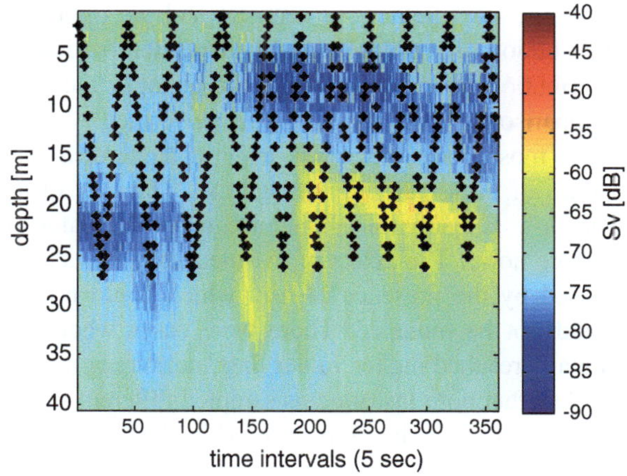

Fig. 7 Echogram with LOPC towing route during the undulating mode (black asterisks); Hornsund, 26 July 2013 21:26 UTC

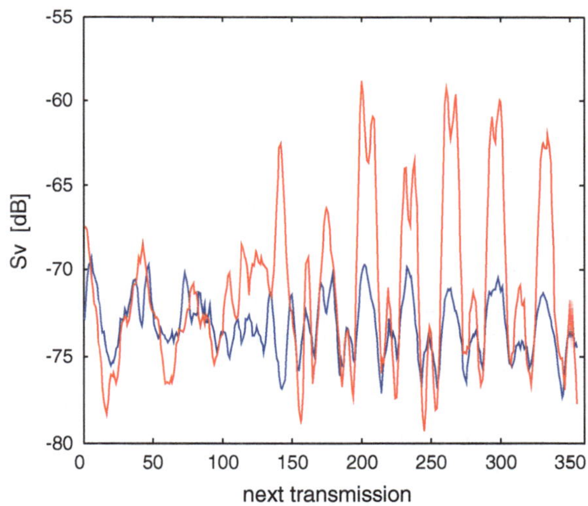

Fig. 8 A comparison of measured (*red*) and modeled (*blue*) values of backscattering strength averaged over 1 m depth layer and 10 transmissions; Hornsund, 26 July 2013 21:26 UTC

5 Discussion

The main goal of this study was to conduct and compare synchronous acoustical and optical measurements. A comparison of different methods is required to assess more thorough information, as each sampling tool suffers its own inefficiencies. Such concurrent measurements have a great potential to describe zooplankton assemblages over large scales and have advantage over single method studies. Acoustic sounding delivers information on distribution of scatterers in the water column, but exact sizes and abundance remain unknown. LOPC measurements produce data on size spectra of particles suspended in the water. Additional measurements of environmental parameters (e.g. temperature, salinity) are helpful in discerning water masses of different properties.

To assess whether the acoustic and the optical method of investigation of the distribution of zooplankton and suspended particles give similar results in terms of distribution in space, as well as to quantify the concentration of zooplankton, the comparison of both methods was performed.

In most cases there was a good agreement between measured and modeled values of backscattering strength S_V. The lack of agreement was in conditions of very low zooplankton abundance when the values of S_V fell to the level of subsurface noise. The peaks of the measured S_V values visible on Fig. 3 are probably caused by the presence of fish, which could not be detected by LOPC and thus could not be modeled. There were cases when the modeled results of S_V occasionally reached higher values then measured ones (Fig. 3). This could be caused by the fact that both instruments sampled different water volumes. Acoustic transducer was mounted to the ship hull, while LOPC was hauled alongside the ship—the particles detected at a given moment by optical sensor could have been omitted by acoustic sounding. The other reason is that backscattering strength is very sensitive to changes in density and sound speed contrasts, which in turn depend on size and properties of each zooplankton individual. Acoustic model assumed homogenous contrasts for whole collection of zooplankton individuals, which was obviously a simplification.

Except the occasional fish presence, the modeled values were also much lower than the measured ones in the case of hydrological front (Fig. 5). LOPC measurements detected high abundance of suspended matter in the colder waters (Fig. 6). Elevated levels of acoustic backscatter were also recorded in the same fragment of the study section (Fig. 7). A comparison of measured and modeled values of backscattering strength indicates that in the glacial waters there is lack of agreement between two methods (Fig. 8). This could suggests that the density and sound speed contrasts used in the model were too low. With use of environmental properties, it could be assumed that the detected suspended matter is most likely mineral of glacial origin and thus having different properties than zooplankton assemblages (i.e. higher contrasts giving higher backscattering strength S_V).

The combined use of acoustic and optical techniques improves the resolution of zooplankton measurements, while additional sensors enable insights into the dynamics of the ecosystem, including environmental forces that regulate zooplankton communities.

6 Conclusions

Acoustical and optical methods supplemented by environmental parameters measurements can identify water masses of different origin. They can also detect zooplankton distribution in time and space as well as differences in zooplankton abundance and size structure.

Backscattering strength measured by echosounder and predicted by "high-pass" model using zooplankton size distribution provided by the LOPC agrees very well, excluding cases of presence of fish and high amounts of suspended matter from glaciers.

Acoustical estimates of the theoretical backscattering strength calculated by the model are very sensitive to changes in density and sound speed contrasts.

In spite of differences in the volume of the water sampled by the acoustical and optical methods, there is a significant, positive correlation between the volume backscattering strength determined by the model and measured by the echosounder.

Acknowledgments We would like to thank Emilia Trudnowska for her assistance in preparing this paper. This work was supported by a Growing of the Arctic Marine Ecosystem (GAME) project financed from the Polish National Science Centre funds under the no. DEC-2012/04/A/NZ8/00661 and by an Acoustical estimation of the abundance and spatio-temporal distributions of the Baltic zooplankton—ZODIAC project financed from the National Science Centre funds under the no. DEC-2013/09/N/ST10/04177.

References

Anderson VC (1950) Sound scattering from a fluid sphere. J Acoust Soc Am 22:426–431

Chu D, Stanton TK, Wiebe PH (1992) Frequency dependence of sound backscattering from live individual zooplankton. ICES J Mar Sci 49:97–106

Johnson RK (1977) Sound scattering from a fluid sphere revisited. J Acoust Soc Am 62:375–377

Kogeler JW, Falk-Petersen S, Kristensen A, Pettersen F, Dalen J (1987) Density- and sound speed contrasts in Sub-Arctic zooplankton. Polar Biol 7:231–235

Krupica KL, Sprules WG, Herman AW (2012) The utility of body size indices derived from optical plankton counter data for the characterization of marine zooplankton assemblages. Cont Shelf Res 36:29–40

Medwin H, Clay CS (1998) Fundamentals of acoustical oceanography. Academic Press, San Diego, p 712

Omori M, Hamner WM (1982) Patchy distribution of zooplankton: behavior, population assessment and sampling problems. Mar Biol 72:193–200

Rayleigh JWS (1945) Theory of sound. Dover, New York

Stanton TK (1989) Simple approximate formulas for backscattering of sound by spherical and elongated objects. J Acoust Soc Am 86(4):1499–1510

Stanton TK, Chu D (2000) Review and recommendations for the modelling of acoustic scattering by fluid-like elongated zooplankton: euphausiids and copepods. ICES J Mar Sci 57:793–807

Trudnowska E, Szczucka J, Hoppe L, Boehnke R, Hop H, Blachowiak-Samolyk K (2012) Multidimensional zooplankton observations on the northern West Spitsbergen Shelf. J Mar Syst 98–99:18–25

Wade IP, Heywood KJ (2001) Acoustic backscatter observations of zooplankton abundance and behaviour and the influence of oceanic fronts in the northeast Atlantic. Deep-Sea Res II Top Stud Oceanogr 48:899–924

Submarine Groundwater Discharge to the Bay of Puck, Southern Baltic Sea and Its Possible Changes with Regard to Predicted Climate Changes

Beata Szymczycha

Abstract The climate change is an ongoing phenomenon causing numerous environmental problems, including modifications of the already seriously influenced by anthropogenic activity hydrological cycle. Estimating the climate change influence on groundwater is challenging because climate change can modify hydrological processes and groundwater resources directly and indirectly. Under the climate scenarios for the southern Baltic, precipitation is projected to increase in the entire Baltic Sea watershed in winter, while in summer increase of precipitation is mainly projected in the northern part of the basin. Thus, the precipitation will impact the groundwater discharge to the sea (SGD). Consequently, the already substantial SGD to the Bay of Puck, southern Baltic Sea can increase. Not only the additional amount of water will enter the marine environment by means of SGD but also significant load of chemical substances.

Keywords Water resources · Global change · Adaptation

1 Introduction

The climate change is an ongoing phenomenon causing numerous environmental problems, including modifications of the hydrological cycle which is already seriously influenced by anthropogenic activity (Dragoni and Sukhija 2008). Climate change affects water resources (e.g. the quality and quantity of groundwater) with wide-ranging consequences for society and ecosystems (Zekster and Loaiciga 1993; Bates et al. 2008; Green et al. 2011).

Estimating the climate change influence on groundwater is challenging because climate change can affect hydrological processes and groundwater resources directly and indirectly (Dettinger and Earman 2007). Moreover, there is a lack of

B. Szymczycha (✉)
Institute of Oceanology Polish Academy of Sciences, Sopot, Poland
e-mail: beat.sz@iopan.gda.pl

T. Zielinski et al. (eds.), *Impact of Climate Changes on Marine Environments*,
GeoPlanet: Earth and Planetary Sciences, DOI 10.1007/978-3-319-14283-8_6

studies on the magnitude and direction of groundwater change due to climate change (Kundzewicz et al. 2007).

The objective of this study is to provide an overview of the research on groundwater discharge to the Bay of Puck, southern Baltic Sea and its changes in relation to predicted climate change.

1.1 Predictions Regarding Climate Change

Climate change can be described as alterations in the global or local climate characteristics and can result in both natural and anthropogenic influences on the hydrological cycle (Green et al. 2011). The increase of greenhouse gases concentration in the atmosphere is assumed to cause the most of the contemporary climate warming while the global atmospheric CO_2 concentration is an important indicator of greenhouse gases (Petit et al. 1999). The increase of CO_2 concentration has been measured atop Mauna Loa, Hawaii at the National Centre for Environmental Protection (NCEP) (e.g. Keeling et al. 1976, 2004; Thoning et al. 1989; Green et al. 2011). The data collected there indicate that CO_2 concentration and its rate of change have been raising continuously since 1958 (the starting year of the measurements) (Fig. 1).

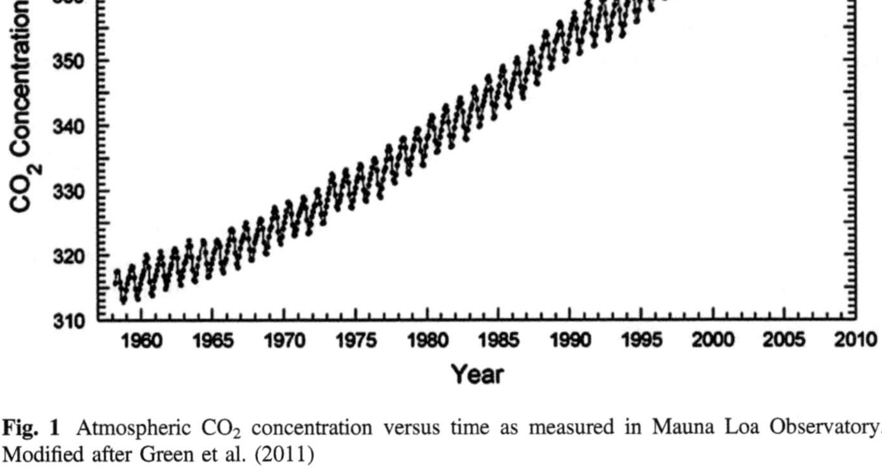

Fig. 1 Atmospheric CO_2 concentration versus time as measured in Mauna Loa Observatory. Modified after Green et al. (2011)

The projections of climate change are based on coupled global climate models, which include the circulation of the atmosphere and the oceans, climate characteristics (rainfall, temperature, radiation, etc.) and different scenarios for increases of the greenhouse gases concentration in the atmosphere (Kløve et al. 2013). The proposed scenarios vary on a constant increase of the greenhouse gases for the next 100 years (scenario A2 of IPCC) to a decrease in emissions (scenario B1 of IPCC) or predictions in between. Therefore, projections from IPCC present significant global warming and alterations in frequency and amount of precipitation from year 2000 to 2100 (IPPC 2007). The global mean surface temperature is estimated to have increased by 0.6 ± 0.2 °C since 1861, and predicts an increase of 2–4 °C over the next 100 years. Temperature growth also distresses the hydrologic cycle by directly increasing evaporation of available surface water and vegetation transpiration. Consequently, these changes can impact amounts of precipitation, timings and intensity rates, and indirectly impact the flux and retention of water in surface and subsurface reservoirs (i.e. lakes, soil moisture, groundwater).

1.2 Possible Climate Change Impact on Groundwater Discharge

Groundwater discharge is commonly known as any outflow of water through the sea floor (Burnett et al. 2006). Groundwater depletion appears when rate of groundwater recharge is smaller than rate of discharge. Unfortunately, groundwater depletion has already affected large regions in many countries around the world as a result of direct and indirect climate change and human activities. The most important one of these is groundwater extraction for irrigated agriculture and/or urban centers (Bouraoui et al. 1999; Brouyere et al. 2004; Alley 2007; Dams et al. 2007; Hsu et al. 2007; Moustadraf et al. 2008). Changing global groundwater discharge has even contributed to sea-level increase over the past century which could have been even greater if significant quantities of water had not been conducted into aquifers by irrigation return-flow (Sahagian et al. 1994).

According to results of climate change modelling some regions can receive more precipitation resulting in increasing groundwater discharge. Submarine groundwater discharge or net groundwater discharge is a substantial component of the global hydrologic cycle, it can be used as a source of fresh water for human needs (Taniguchi 2000; Burnett et al. 2006). Estimating submarine groundwater discharge and its biogeochemical effects on the marine environment has important implications for understanding climate-change impact on marine processes (Windom et al. 2006). There are examples of high dissolved nitrogen-phosphorous ratios in SGD relative to surface water that can drive the coastal oceans toward phosphorous limitation within next decades which can result in changing the present nitrogen-limited coastal primary production (Slomp and Van Cappellen 2004; Taniguchi et al. 2008).

2 The Characterization of the Baltic Sea and Its Future Climate

2.1 The Baltic Sea

The Baltic Sea is located in the Northern Europe and is almost entirely surrounded by land. The eastern, southern and western parts of the Baltic Sea border with the European continent (Russia, Estonia, Latvia, Lithuania, Poland, Germany, Denmark), while the northern part is limited by Scandinavian Peninsula (Finland and Sweden). The only connection to the North Sea is through the Kattegat and the Skagerrak. The Baltic Sea is one of the largest brackish seas in the world with a total surface area of 415,240 km^2 (including the Danish Sounds and the Kattegat) and the corresponding volume of water 21,706 km^3 (Voipio 1981). Generally, the Baltic Sea is separated into several basins by elevations and thresholds: the Bothnian Bay, the Bothnian Sea, the Gulf of Finland, the Gulf of Riga, the Baltic Proper, the Danish Sounds and the Kattegat.

The total drainage area of the Baltic Sea is 1,729,000 km^2. Thus, it is more than four times larger than the sea itself. The major Baltic Sea rivers are Neva, Vistula, Oder, Neman, and Daugava which drain the southern and south-eastern lowlands. River water plays an important role as a component of the water balance, and forming and modifying the salinity of the Baltic Sea water. The Baltic Sea catchment includes also a large number of lakes of postglacial origin (e.g. Ladoga and Onega).

The Baltic Sea is impacted by substantial fresh water inputs and the sporadic inputs of saline, North Sea water (Voipio 1981). The salinity in the Baltic Sea varies depending on the depth and location. Usually, the surface water and deeper water domains are separated by stable halocline. In the rivers mouths the surface water salinity can reach zero while in the Bothnian Bay, Gulf of Finland and Gulf of Riga the salinity increase to 3–5. In the Baltic Proper surface water salinity equals to some 7.5. Characteristic of the Baltic Sea is a layered distribution of salinity in the vertical profile. Within the profile two layers can be distinguished: an upper, less saline isohaline layer and lower layer of more salty water. The layers are separated by a halocline. The depth of the halocline is spatially variable.

The Baltic Sea fills a depression in the continental crust below present sea level (Uścinowicz and Miotk-Szpiganowicz 2011). There are several forces like erosion, tectonic and glacio-isostatic processes that influence the geology, morphology and bathymetry of the Baltic Sea. The Baltic Sea continental catchment area is characterized by lowlands covered by Quaternary deposits (Uścinowicz and Miotk-Szpiganowicz 2011).

2.2 Submarine Groundwater Discharge in the Baltic Sea

Submarine groundwater discharge (SGD) is one of the water pathways connecting land and ocean in the global water cycle. It has been recognized as an important factor influencing coastal zones (Burnett et al. 2006).

Initially, investigations concerning groundwater system in the Baltic Sea Region were limited to seawater intrusion into the Baltic Sea coast (e.g. Agopsowicz and Pazdro 1964; Kolago 1964; Lidzbarski 2011). Later, scientists focused on groundwater discharge into the Baltic Sea and hydrogeological conditions of the sea bottom in different parts of the Baltic Sea. Thus, the lake-lands on morainic uplands located along the southern Baltic Sea coast were identified as major SGD sources. Assessments of the groundwater discharge to the Baltic Sea and defined groundwater discharge areas are presented in Table 1.

Pietrucień (1983) and Peltonen (2002) estimated the groundwater discharge to the Baltic Sea to be 0.78 and 4.4 km^3 $year^{-1}$, respectively. However, the calculation made by Pietrucień (1983) did not include deeper, Paleogene, Cretacaceus and Jurrasic aquifers. Peltonen (2002) calculated groundwater discharge along the German and Polish coasts to be 1.4 km^3 $year^{-1}$. This estimation was based on regional studies in Poland. Kryza and Kryza (2006) made a model estimation of the direct groundwater inflow to the Baltic Sea from Polish territory. The result was one order of magnitude smaller than that calculated by Peltonen (2002) and equaled 0.2 km^3 $year^{-1}$. The possible explanation of smaller groundwater discharge rate calculated by Kryza and Kryza (2006) compared to the one made by Pelotnen (2002) is that Kryza and Kryza (2006) assumed that groundwater seeps already at the shoreline. In the Gulf of Finland total amount of groundwater discharge was calculated to be 0.6 km^3 $year^{-1}$ (Viventsowa and Voronow 2003). Lidzbarski (2011) estimated the approximate groundwater discharge into the Gulf of Gdańsk to be 0.07 km^3 $year^{-1}$. Piekarek-Jankowska (1994) calculated groundwater discharge

Table 1 The groundwater discharge to the Baltic Sea and selected groundwater discharge areas

Study area	Groundwater discharge		References
	km^3 $year^{-1}$	m^3 h^{-1}	
The Baltic Sea	0.8	89,041	Pietrucień (1983)
The Baltic Sea	4.4	502,283	Peltonen (2002)
The Polish and German Coast[a]	1.4	162,000	Peltonen (2002)
The Polish Coast	0.2	16,568	Kryza and Kryza (2006)
The Gulf of Finland	0.6	68,493	Viventsova and Voronow (2003)
The Gulf of Gdańsk	0.07	8,000	Lidzbarski (2011)
The Bay of Puck	0.03	3,424	Piekarek-Jankowska (1994)
The Eckernfőrde Bay	0.16	18,535	Schlüter et al. (2004)

[a] The estimation based on the local observations

from the Quaternary, Neogene, Paleogene and Upper Cretaceous aquifers to the Bay of Puck to be 0.03 km^3 $year^{-1}$. Schlüter et al. (2004) estimated that the average discharge rate to the Eckernfőrde Bay was equal to 0.16 km^3 $year^{-1}$.

2.3 Predictions of Future Baltic Sea Climate Change

The climate change in the Baltic Sea region has been a subject of several international projects. As a result the knowledge of the past changes in climate have progressed in the past two decades. Moreover, increasing effort is being paid to comprehend and predict future climate change of the Baltic Sea region.

The assessment of the future climate change is based on general circulation models which are downscaled by many methods (etc. statistical downscaling, dynamical downscaling) or on regional climate models. Consequently the projected climate changes with regard to different models have different uncertainties. Nonetheless, model simulations project an increase in temperature in the Baltic Sea region similarly to an increase of the global mean temperature. According to recent findings (Baltic Sea Environment Proceedings No. 137 2013) the largest warming was projected in the north in winter. However, warm extremes in summer are also expected to appear more often. It has been measured that surface waters in the Baltic Sea have warmed in all seasons by up to 1 °C/decade since 1985. In a future climate, the summer sea-surface temperature growth is likely to be about 2 °C in the southern parts of the Baltic Sea and about 4 °C in the northern parts till 2100. Moreover, sea-ice cover scenario simulations indicate its decrease in the Baltic Sea in the future.

The precipitation in the Baltic Sea region during the past 100 years has not shown a stable trend. However, a tendency of the increasing precipitation in winter and spring has detected during the second half of the 20th century. Precipitation impacts river runoff. Thus, river runoff over the past century have shown a decrease in annual discharges from southern part of the catchment. This indicates that the southern regions of the Baltic Sea Basin may become drier with rising air temperatures. In contrast, trends in the north and around the Gulf of Finland indicate increased annual stream flows under warmer temperatures (BSEP 137).

The predictions of future precipitation indicate an increase in the entire Baltic Sea region during winter, while in the summer increases in precipitation are mainly projected only for the northern half of the basin.

The models simulations concerning future sea-level rise are still uncertain although the high-end scenarios project from 0.6 to 1.1 m sea-level rise for the next century for the Baltic Sea.

3 Possible Climate Change Impact on Groundwater Discharge to the Bay of Puck, Southern Baltic Sea

3.1 Characterization of the Bay of Puck

The Bay of Puck is located in the southern part of the Baltic Sea (eastern part of the Gulf of Gdańsk). A narrow, sandy spit called the Hel Peninsula, which is 36 km long, separates the bay from the open waters of the Baltic Sea. The total area of the Bay of Puck equals 359.2 km^2. The catchment area is nearly three times larger and equals 908.8 km^2. The major rivers flowing to the Bay of Puck are Reda, Gizdepka, Płutnica, Zagórska Struga and Chylonka. The total river inflow to the Bay of Puck is equal to 0.25 km^3 year^{-1} while the average precipitation equals 0.20 km^3 year^{-1} (Cyberski and Szefer 1993). The Bay of Puck consists of an outer southeastern region with an average depth of 20.5 m, and an inner northwestern region with the average depth of 3.1 m. The basin slopes downward in a northeasterly direction reaching the depth of 54 m near the tip of the Hel Peninsula (Cyberski and Szefer 1993). An average salinity in the Bay of Puck is in the range from 7.00 to 7.65 (Nowacki 1993).

The geomorphology of the Bay of Puck is complex. A hydrogeological section of the Bay of Puck is presented on Fig. 2. The Quaternary sediments in the Bay of Puck are some 25 m thick. The Cretaceus formation lies at depth from 108–135 m Seismic–acoustic investigations of the study area have imaged permeable layers of Holocene to Pleistocene sands and silts and underlying Tertiary silt layers (Piekarek-Jankowska 1994). The aquifers formed in the Tertiary forms of the Oligocene and Miocene sandy, sediments. The Hel Peninsula developed during the Pleistocene and Holocene epochs (Piekarek-Jankowska 1994). Geomorphic landforms surrounding the Bay of Puck consist of wave-dominated sedimentary plains and dune

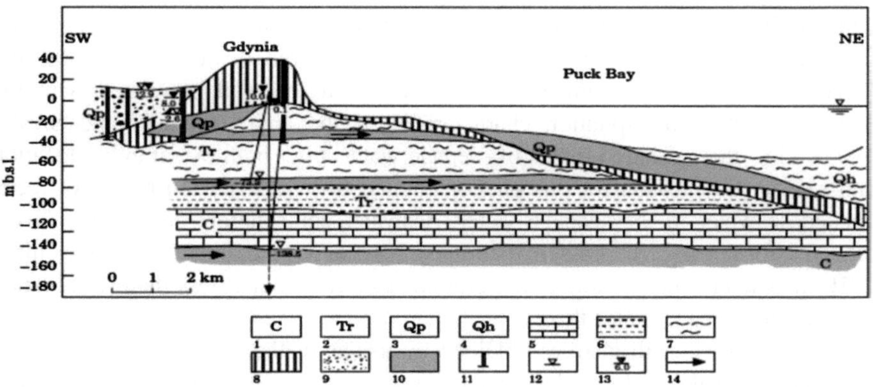

Fig. 2 A Bay of Puck cross-section presenting hydrogeological conditions of the sediments (Falkowska and Piekarek-Jankowska 1999). *1* Cretaceus, *2* Tertiary, *3* Quatenary-Pleistocene, *4* Quatenary-Holocene, *5* marl, *6* clay, *7* slit, *8* boulder, *9* clay, *10* sand, *11* well, *12* *top* of a aquifer, *13* piezometric groundwater level, *14* direction of groundwater flow

deposits forming in micro tidal zones. Coastal erosion is the dominant source of sediments within the study area. Waves, storm surges, currents and winds driven erosion are reason for transport, accumulation and redeposition of sediments in the coastal zone (Piekarek-Jankowska 1994).

3.2 Submarine Groundwater Discharge to the Bay of Puck

The groundwater discharge to the Bay of Puck was mainly observed as salinity changes of seafloor water level. It is said that about 50 % of sediments in the Bay of Puck are impacted by groundwater discharge (Piekarek-Jankowska 1994). The identified groundwater impacted areas are located in the inner part of the Bay of Puck (the western part of the inner Bay of Puck from Płutnica river estuary to the mouth of the Reda valley and in the outer part of the Bay of Puck (in the middle part of the reservoir). Piekarek-Jankowska (1994) calculated groundwater discharge from the Quaternary, Neogene, Paleogene and Upper Cretaceous aquifers to the Bay of Puck to be 0.03 km^3 $year^{-1}$. Another study calculating groundwater discharge to the Bay of Puck was made by Szymczycha et al. (2012). In the study—seepage fluxes, salinity rates and the end-member method were used to calculate groundwater discharge into the study area. This combination uses sensitive and precise measurements of end-member components to estimate groundwater and seawater contributions to seepage fluxes (Szymczycha et al. 2012). Measurements were repeated three times per each seepage meter location during four sampling campaigns: September 2009; November 2009; February 2010 and May 2010 (Pempkowiak et al. 2010). The obtained results are presented in Table 2. Variations on the results obtained with a single seepage meter did not exceed 5 %. However, the differences in seepage water fluxes between the locations (S1, S2, S3) during each sampling campaign were significant. In February 2010 the relative standard deviation (RSD) of the calculated average seepage water flux reached the highest equal to 70 % value. Moreover, the data collected indicate that fluxes in February 2010 and May 2010 were lower than fluxes measured in September 2009 and November 2009. The average SGD turned out to be well correlated with the average monthly precipitation characteristic of the area (Cyberski and Szefler 1993). Thus the measured fluxes of seepage water differ both by sampling location and season. One reason for this phenomenon is the varying contribution of recirculated seawater to the seepage water. Seawater contributions to seepage fluxes ranged between 4 L d^{-1} m^{-2} in February 2010 and 44 L d^{-1} m^{-2} in November 2010. It is interesting to notice that groundwater contributed less than recirculated seawater to seepage water fluxes.

The groundwater discharges obtained by means of the seepage meters method combined with the end-member method were used to calculate the total groundwater discharge rate to the Bay of Puck. The total groundwater drainage area of the Bay of Puck equals 200 km^2—some 56 % of total surface area of the bay

Table 2 Seepage water fluxes and groundwater discharge to the study area

Date	Location	Salinity of seepage water	Fluxes ($L\ d^{-1}\ m^{-2}$)		Groundwater discharge seasonal average ± SD
			Seepage water	Groundwater	
September 2009	S1	5.6 ± 0.5	84.0 ± 3.6	23.7 ± 1.1	25.8 ± 2.4
	S2	5.2 ± 0.6	89.7 ± 3.4	25.6 ± 0.5	
	S3	4.1 ± 0.4	64.4 ± 2.4	28.5 ± 1.1	
November 2009	S1	6.5 ± 0.7	187.1 ± 7.0	18.7 ± 0.5	18.4 ± 5.4
	S2	6.1 ± 0.6	150.4 ± 5.7	23.6 ± 0.7	
	S3	6.3 ± 0.7	99.9 ± 3.8	15.8 ± 0.9	
February 2010	S1	5.6 ± 0.6	10.2 ± 0.4	2.3 ± 0.1	3.0 ± 2.1
	S2	4.7 ± 0.5	15.1 ± 0.6	5.4 ± 0.2	
	S3	5.8 ± 0.6	7.1 ± 0.3	1.4 ± 0.1	
May 2010	S1	3.7 ± 0.4	7.3 ± 0.3	3.7 ± 0.1	3.6 ± 0.1
	S2	6.5 ± 0.7	36.7 ± 1.4	3.7 ± 0.1	
	S3	5.9 ± 0.6	19.3 ± 0.7	3.6 ± 0.1	

Seepage water samples were collected and salinity was measured three times per each sampling location (S1, S2, S3) during four sampling campaigns: September 2009; November 2009; February 2010 and May 2010

(359.2 km^2) (Piekarek-Jankowska 1994). The average groundwater discharge is equal to 0.9 ± 0.8 $km^3\ year^{-1}$. Piekarek-Jankowska (1994) calculated that total groundwater discharge to the Bay of Puck equals 0.03 $km^3\ year^{-1}$. The result obtained during this study is 30 times higher than that obtained by Piekarek-Jankowska (1994). One reason for this is that groundwater is discharged irregularly. There are subareas characterized by lower and higher SGDs. However, the Bay of Puck seafloor is characterized by areas of seepage water discharge and areas with no discharge, thus using the total surface area of the Bay of Puck for calculating SGD could have led to overestimation. It seems that the study area has larger SGD fluxes than the remaining position of the Bay of Puck.

Nutrients fluxes via SGD to the Bay of Puck were calculated using the obtained groundwater fluxes and nutrient concentrations measured in groundwater (Szymczycha et al. 2012). The loads of ammonium and phosphate were significantly higher than the combined nitrate and nitrite loads. The high fluxes of ammonium and phosphate can be explained by high concentration of ammonium and phosphate in groundwater. Generally, groundwater is enriched with nitrate and phosphate (Moore 2010). In this study, due to the anoxic conditions in the sediments, nitrate and nitrite are reduced, and consequently low loads of nitrate and nitrite are delivered via SGD to the study area.

Nutrients fluxes to the Bay of Puck (both inorganic nitrogen—DIN, and phosphates—PO_4^{3-}) were calculated using literature groundwater flux to the Bay of Puck (Piekarek-Jankowska 1994) and the measured yearly averages of nutrients concentrations (Szymczycha et al. 2012). The calculated fluxes equaled to 30 ± 11 t N-DIN $year^{-1}$ and 60 ± 5 t P-PO_4^{3-} $year^{-1}$. Phosphate fluxes via SGD are significant. They are approximately 3 times larger than the atmospheric deposition and comparable to phosphate fluxes via rivers and point sources.

3.3 Predicted Groundwater Discharge and Accompanying Nutrients Fluxes to the Bay of Puck, Southern Baltic Sea

Submarine groundwater discharge has been recognized as an important factor influencing coastal zones (Burnett et al. 2006; Moore 2010). However, there are locations where an increased groundwater usage has lowered potentiometric heads in coastal aquifers and caused infiltration of sea water into the groundwater formations (Green et al. 2011). The result has been an increased rate of salinization of coastal potable water. Sea level rise, whether natural or induced, has also caused increased salinization of groundwater. These changes have been taking place worldwide due to increased demands for fresh water in coastal areas. Consequently, due to expanded anthropogenic pressure, groundwater discharge can decrease (Moore 2010). Despite this raising inland expansion can lead to larger SGD fluxes of nutrients, carbon and metals because the biogeochemical reactions that affect their concentrations can operate over greater spatial scales and impact aquifers that have been in contact with sea water for thousands of year (Moore 2010).

Under the climate scenarios for the southern Baltic, precipitation is projected to increase in the entire Baltic Sea region during winter, while in the summer increases in precipitation are mainly projected for the northern half of the basin. For the southern part of the Baltic Sea, there is a large spread between the different models indicating both increases and decreases and thus no clear change in precipitation can be projected (BSEP 137 2013). Szymczycha et al. (2012) indicated that average SGD turned out to be well correlated with the average monthly precipitation characteristic of the area. Thus, the precipitation actually impacts the groundwater discharge. According to Szymczycha et al. (2012) in winter 2009 the groundwater discharge was the smallest and equaled to 3.0 ± 2.1 L d^{-1} m^{-2}. Thus the predicted increase of precipitation for the Baltic Sea region during winter time can also increase the groundwater fluxes. Consequently, the annul SGD to the Bay of Puck, southern Baltic Sea will increase. Not only the additional flux of water will enter the marine environment but also significant amount of nutrients, organic matters and heavy metals (Szymczycha et al. 2012, 2013, 2014). In summer submarine groundwater discharge can vary but still the loads of chemical compounds in groundwater can increase due to anthropogenic pressure.

4 Conclusions

SGD will increase and probably more nutrients, metals, etc. will enter the Bay of Puck. The increase of the phosphate fluxes via SGD will further enhance the nitrate-limitation of the coastal primary production (Korzeniewski 1993). The projected SGD contribution to the overall phosphate input into the Bay of Puck will be significant. It can be assumed that SGD impact on overall phosphate concentration in the Baltic Sea will be also noteworthy. Thus, further studies are critical to fully understand the apparently significant role of SGD and climate change on the Baltic Sea marine ecosystem.

Acknowledgments The study reports the results obtained within research project 2012/05/N/ST10/02761 sponsored by the Polish Ministry of Science and Higher Education and as a part of the Institute of Oceanology Polish Academy of Sciences statutory activities.

References

Alley WM (2007) Flow and storage in groundwater systems. Science 296:1985–1990

Agopsowicz T, Pazdro Z (1964) Zasolenie wód kredowych na Niżu Polskim. Zeszyty naukowe Politechniki Gdańskiej 6:151–162

Bates B, Kundzewicz ZW, Wu S, Palutikof JP (2008) Climate change and water. Technical paper VI of the intergovernmental panel on climate change. Intergovernmental Panel on Climate Change Secretariat, Geneva, 210 pp

Bouraoui F, Vachaud G, Li LZX, Le Treut H, Chen T (1999) Evaluation of the impact of climate changes on water storage and groundwater recharge at the watershed scale. Clim Dyn 15 (2):153–161

Brouyere S, Carabin G, Dassargues A (2004) Climate change impacts on groundwater resources: modelled deficits in a chalky aquifer, Geer Basin, Belgium. Hydrogeol J 12(2):123–134

Baltic Sea Environment Proceedings No. 137 (2013) Climate change in the Baltic Sea Area. HELCOM thematic assessment in 2013. Helsinki Commission Baltic Marine Environment Protection Commission

Burnett WC, Aggarwal PK, Aureli A, Bokuniewicz H, Cable JE, Charette MA, Kontar E, Krupa S, Kulkarni KM, Loveless A, Moore WS, Oberdorfer JA, Oliveira J, Ozyurt N, Povinec P, Privitera AMG, Rajar R, Ramessur RT, Scholten J, Stieglitz T, Taniguchi M, Turner JV (2006) Quantifying submarine groundwater discharge in the coastal zone via multiple methods. Sci Total Environ 367(2–3):498–543

Cyberski J, Szefler K (1993) Klimat Zatoki i jej zlewiska: Zatoka Pucka. Edited by Korzeniewski K. Fundacja Rozwoju Uniwersytetu Gdańskiego, Gdańsk, Poland, pp 14–39

Dams J, Woldeamlak ST, Batelaan O (2007) Forecasting land-use change and its impact on the groundwater system of the Kleine Nete catchment, Belgium. Hydrol Earth Syst Sci Discuss (HESS-D) 4(6):4265–4295

Dettinger MD, Earman S (2007) Western ground water and climate change—pivotal to supply sustainability or vulnerable in its own right? Ground Water 4(1):4–5

Dragoni W, Sukhija BS (2008) Climate change and groundwater: a short review. Geol Soc Lond Spec Publ 288:1–12. doi:10.1122/SP288.1

Falkowska L, Piekarek-Jankowska H (1999) The submarine seepage of the fresh water: disturbance in hydrological chemical structure of the water column in the Gdansk Deep. J Mar Sci 56:153–160

Green TR, Taniguchi M, Kooi H, Gurdak JJ, Allen DM, Hiscock KM, Treidel H, Aureli A (2011) Beneath the surface of global change: impacts of climate change on groundwater. J Hydrol 405:532–560. doi:10.1016/j.jhydrol.2011.05.002

Hsu KC, Wang CH, Chen KC, Chen CT, Ma KW (2007) Climate-induced hydrological impacts on the groundwater system of the Pingtung Plain, Taiwan. Hydrogeol J 15(5):903–913

IPCC 2007 (2007) Climate change 2007: the physical science basis. Contribution of working group I to the fourth assessment report of the intergovernmental panel on climate change. In: Solomon S et al (eds) Cambridge University Press, Cambridge, UK, and New York, USA

Keeling CD, Bacastow RB, Bainbridge AE (1976) Atmospheric carbon dioxide variations at Mauna Loa observatory, Hawaii. TELLUS 28(6):538–551

Keeling CD, Brix H, Gruber N (2004) Seasonal and long-term dynamics of the upper ocean carbon cycle at station ALOHA near Hawaii. Global Biogeochem Cycles 18(4):1–26

Kløve B, Ala-Aho P, Bertrand G, Gurdak JJ, Kupferberger H, Kværner J, Muotka T, Mykra H, Preda E, Rossi P, Uvo BC, Velasco E, Pulido-Velazquea M (2013) Climate change impacts on groundwater and dependent ecosystems. J Hydrol (in press)

Kolago C (1964) Wody mineralne województwa szczecińskiego i perspektywy ich wykorzystania. Przegląd Zachodniopomorski 5:65–85

Korzeniewski K (1993) Zatoka Pucka. Instytut Oceanologii Uniwersytetu Gdanskiego, Gdynia

Kryza J, Kryza H (2006) The analytic and model estimation of the direct groundwater flow to Baltic Sea on the territory of Poland. Geologos 10:154–165

Kundzewicz ZW, Mata LJ, Arnell NW, Doll P, Kabat P, Jimenez B, Miller KA, Oki T, Sen Z, Shiklomanov IA (2007) Freshwater resources and their management. In: Parry ML, Canziani OF, Palutikof JP, van der Linden PJ, Hanson CE (eds) Climate change 2007: impacts adaptation and vulnerability. Cambridge University Press, Cambridge, pp 173–210

Lidzbarski M (2011) Groundwater discharge in the Baltic Sea Basin: geochemistry of Baltic Sea surface and sediments. Edited by Uścinowicz Sz, 2011. Polish Geological Institute-National Research Institute, Warsaw, pp 138–145

Moore WS (2010) The effect of submarine groundwater discharge on the ocean. Annu Rev Mar Sci 2:59–88

Moustadraf J, Razack M, Sinan M (2008) Evaluation of the impacts of climate changes on the coastal Chaouia aquifer, Morocco, using numerical modeling. Hydrogeol J 16(7):1411–1426

Nowacki J (1993) Termika, zasolenie i gęstość wody: Zatoka Pucka. In: Korzeniewski K (ed). Fundacja Rozwoju Uniwersytetu Gdańskiego, Gdańsk, Poland, pp 79–112

Pempkowiak J, Szymczycha B, Kotwicki L (2010) Submarine groundwater discharge (SGD) to the Baltic Sea. Rocznik Ochrony Środowiska 12:17–32

Peltonen K (2002) Direct groundwater inflow to the Baltic Sea. TemaNord, Nordic Councils of Ministers, Copenhagen, Holand, 79 pp

Petit JR, Jouzel J, Raynaud D, Barkov NI, Barnola JM, Basile I, Bender M, Chappellaz J, Davis M, Delaygue G, Delmotte M, Kotlyakov VM, Legrand M, Lipenkov VY, Lorius C, Pepin L, Ritz C, Saltzman E, Stievenard M (1999) Climate and atmospheric history of the past 420,000 years from the Vostok ice core, Antarctica. Nature 399(6735):429–436

Piekarek-Jankowska H (1994) Zatoka Pucka jako obszar drenażu wód podziemnych. Wydawnictwo Uniwersytetu Gdańskiego, Gdańsk, Poland

Pietrucień Cz (1983) Regionalne zróżnicowanie warunków dynamicznych i hydrodynamicznych wód podziemnych w stresie brzegowej południowego i wschodniego Bałtyku. Turuń, Poland, p 71

Sahagian DL, Schwartz FW, Jacobs DK (1994) Direct anthropogenic contributions to sea level rise in the twentieth century. Nature 367:54–57

Schlüter M, Sauter EJ, Andersen CA, Dahlgaard H, Dando PR (2004) Spatial distribution and budget for submarine groundwater discharge in Eckernförde Bay (Western Baltic Sea). Limnol Oceanogr 49:157–167

Slomp CP, Van Cappellen P (2004) Nutrient inputs to the coastal ocean through submarine groundwater discharge: controls and potential impact. J Hydrol 295(1–4):64–86

Szymczycha B, Vogler S, Pempkowiak J (2012) Nutrient fluxes via submarine groundwater discharge to the Bay of Puck, southern Baltic Sea. J Total Environ 438:86–93

Szymczycha B, Miotk M, Pempkowiak J (2013) Submarine groundwater discharge as a source of mercury in the Bay of Puck, the Southern Baltic Sea. Water Air Soil Pollut 224. doi:10.1007/s11270-013-1542-0

Szymczycha B, Maciejewska A, Winogradow A, Pempkowiak J (2014) Could the submarine groundwater discharge be a significant carbon source to the southern Baltic Sea? Oceanologia 56:327–347

Taniguchi M (2000) Evaluation of the saltwater–groundwater interface from borehole temperature in a coastal region. Geophys Res Lett 27(5):713–716

Taniguchi M, Burnett WC, Ness GD (2008) Integrated research on subsurface environments in Asian urban areas. Sci Total Environ 404(2–3):377–392

Thoning KW, Tans PP, Komhyr WD (1989) Atmospheric carbon dioxide at Mauna Loa observatory. 2. Analysis of the NOAA GMCC data, 1974–1985. J Geophys Res 94(D6):8549–8565

Uścinowicz Sz, Miotk-Szpiganowicz G (2011) The Baltic Sea: location, division and catchment area: geochemistry of Baltic Sea surface and sediments. Edited by Uścinowicz Sz, 2011. Polish Geological Institute-National Research Institute, Warsaw, pp 13–17

Voipio A (1981) The Baltic Sea. Elsevier Scientific Publishing Company, Amsterdam, p 148

Viventsowa EA, Voronow AN (2003) Groundwater discharge to the Gulf of Finland (Baltic Sea): ecological aspects. Environ Ecol 45:221–225

Windom HL, Moore WS, Niencheski LFH, Jahnke RA (2006) Submarine groundwater discharge: a large, previously unrecognized source of dissolved iron to the south Atlantic ocean. Mar Chem 102:252–266

Zektser IS, Loaiciga HA (1993) Groundwater fluxes in the global hydrologic cycle: past, present and future. J Hydrol 144(1–4):405–427

Climate Change Influence on Migration of Contaminants in the Arctic Marine Environment

Anna Pouch and Agata Zaborska

Abstract The Arctic is particularly vulnerable to environmental changes connected to climate change. Most of assessments of the climate change impact to the Arctic environment concentrate on direct effects to the marine and terrestrial ecosystems. There is little understanding of numerous indirect effects of global change and their impact on cycle of different compounds e.g. man-made substances. The global change effects will not always be predictable but may be abrupt. Environmental changes connected to climate change will influence contaminant transport and migration within the Arctic marine ecosystem. Main effects of global change will be visible through changes of large scale contaminant transport pathways e.g. air mass transport, ice transport, marine currents transport and the changes of in situ environmental conditions e.g. changes of pH, temperature, oxygen content. In this review article we describe major environmental factors that may influence global transport of contaminants and migration of contaminants within the arctic ecosystem elements. We also discuss possible further changes in contaminant sources and distribution within the Arctic related to global changes.

1 Introduction

The Arctic plays an important role for the global climate system but simultaneously it is extremely vulnerable to changes of environmental conditions. In the Arctic warming occurs faster than in other parts of the globe and polar regions are predicted to incur some of the most pronounced effects of the climate change (McBean et al. 2004; Johannessen et al. 2004; IPCC 2013). Satellite data show statistically

A. Pouch · A. Zaborska (✉)
Polish Academy of Sciences, Institute of Oceanology, Powstancow Warszawy 55,
81-712 Sopot, Poland
e-mail: agata@iopan.gda.pl

A. Pouch
Centre for Polar Studies, National Leading Research Centre, Bedzinska 60,
41-200 Sosnowiec, Poland

T. Zielinski et al. (eds.), *Impact of Climate Changes on Marine Environments*,
GeoPlanet: Earth and Planetary Sciences, DOI 10.1007/978-3-319-14283-8_7

significant decreases in sea ice area of about 3 % per decade since the late 1970s, accompanied by a 14 % reduction in wintertime multiyear ice (Vavrus and Harrison 2003). Over last century, average sea ice extents declined by 15 % in summer, and 8 % in spring. Reductions in sea ice cover are expected to alter air-sea interactions, surface albedo and salinity gradients, which influence heat transfer, stability of the water column and, ocean circulation patterns (Clarke and Harris 2003). Sea ice restricts exchange processes between ocean and atmosphere. Its presence or absence modifies the albedo of the ocean, as well as the exchange of heat, moisture, and trace gases such as CO_2, between the atmosphere and ocean. In addition to it's climatic role, what is of major importance for the biology of the polar oceans, for the atmospheric chemistry of the polar lower atmosphere, and for the economy and geopolitics of the Arctic region (Abram et al. 2013).

Most of assessments of the climate change in the Arctic concentrate on changes of environmental factors that govern marine and terrestrial ecosystems. There is little understanding of numerous indirect effects of global change and their influence on cycle of different man-made substances. Environmental changes connected to climate change will influence contaminant transport and distribution within the Arctic marine ecosystem (Macdonald et al. 2005). The global change effects will not always be direct and predictable but may be abrupt. Main effects of global change will be visible through changes of large scale contaminant transport pathways e.g. air mass transport, ice transport, marine currents transport and the changes of in situ environmental conditions e.g. changes of pH, temperature, oxygen content. Particularly enhanced contaminant transport from melting permafrost is expected to increase contaminant levels in the marine environment (Rydberg et al. 2010).

The major contaminant groups of growing concern in the Arctic are acidifying gases (SOx) from Eurasian smelters and industry; heavy metals from fossil fuel combustion and mining (Smith et al. 1998; Gobeil et al. 2001) and persistent organic pollutants (POPs) including pesticides used in agriculture (e.g. DDT), and polychlorinated biphenyls (PCBs) leached from electronic transformers (Halsall et al. 2001; Wania and Su 2004; Zaborska et al. 2011). The Arctic region is also especially vulnerable to radioactive contamination originating from global (Karcher et al. 2010; Wit and Muir 2010; Zaborska et al. 2010) and local sources (Pogrebov et al. 1997; Smith et al. 2000).

In this review article we describe major environmental factors that may influence global transport of contaminants and migration of contaminants within the ecosystem elements. We also discuss possible further changes in contaminant sources and distribution within the Arctic related to global changes.

2 Changes in Global Contaminants Transport

Global environmental changes connected to climate change will influence contaminant transport pathways. Changes in intensity of transport and contaminant concentrations discharged to the Arctic are expected. Contaminants may be

Fig. 1 The major pathways of transport of contaminants to the Arctic. Contaminants are mainly transported by air, sea currents, rivers and sea ice (Macdonald et al. 2005)

transported to the Arctic from several remote and local sources by different transport pathways (Fig. 1):

- air transport driven by atmospheric circulation
- ocean transport
- riverine transport
- sea ice transport
- coastal erosion and ground water discharge
- biological transport driven by migrating animals

2.1 The Air Transport

The atmosphere is the fastest and most direct contaminant transport route from distant and local sources to the Arctic (Halsall et al. 2001; Gordeev 2002; Melnikova et al. 2003; Wania and Su 2004). This pathway is particularly efficient for volatile and semi-volatile contaminants as persistent organic pollutants (POPs) and some heavy metals (Hg, Pb). Transport by winds may deliver contaminants from the south to the Arctic within a few days (Halsall et al. 1998). From the atmosphere, the contaminants may enter marine arctic ecosystem via direct air/water/ice exchange but the main process discharging contaminants to the sea surface is wet or dry precipitation. The atmosphere is also important for radionuclides transport that reached higher atmosphere parts during air detonations of nuclear bombs (particularly in 60s and 70s) and still are present in the upper stratosphere (Crane et al. 2000; Efurd et al. 2005).

Atmospheric circulation in the Arctic is driven by Arctic Oscillation (AO) that is correlated with North Atlantic Oscillation (NAO). North Atlantic Oscillation is described as NAO index that is normalized air pressure difference between Iceland and Azores. In the summer, the continental high pressure cells disappear what results in weakening of northward transport from low latitudes. Summer transport accounts in average for only 20 % of the annual south to north air transport (from Norwegian Sea: 10 %, Eastern Europe/Siberia: 5 %, and Bering Sea: 5 %). In the winter, the lower tropospheric circulation is dominated by high pressures over the continents and low pressures over the northern Pacific (Aleutian Low) and Atlantic Oceans (Icelandic Low). Contaminants are transported to the Arctic during the winter by three routes—from the Norwegian Sea (40 %), Eastern Europe/Siberia (15 %), and the Bering Sea (25 %) (Macdonald et al. 2005). Particularly in this season, the cold condensation processes efficiently remove contaminants from the atmosphere to the sea surface.

The Arctic aerosol composition shows an annual variation with maximal concentration values of anthropogenic components during the winter/spring season. The winter phenomenon of "ice crystal haze" was discovered not to be wind-blown dust, but air pollution from the mid-latitudes. Arctic haze is a mixture of aerosols containing acidifying gases (SOx and NOx), coarse particles of soot, heavy metals, polycyclic aromatic hydrocarbons (PAHs), and polychlorinated bifenyls (PCBs). Arctic haze is concentrated in the lower troposphere (up to 3 km altitude) and is most pronounced during the coldest months of the year (from December–April) (Bard 1999).

The atmospheric transport apart from transport from remote globe regions also includes local sources. Two-thirds of the heavy metals in the air in the Arctic originate from industrial activities on the Kola Peninsula, the Norilsk industrial complex, the Pechora Basin, the Urals and the Yakutsk region of the Russian Arctic (Gordeev 2002). Other local sources include for instance mining activities at Svalbard Archipelago and Greenland (Elberling et al. 2002; Dowdall et al. 2004).

The recent warming is consistent with observed changes in atmospheric circulation, particularly the generally positive phase of the North Atlantic/Arctic Oscillation (Vavrus and Harrison 2003). The NAO index generally tend to be more often positive what results in more intense and extending farther north wind transport. The contaminants may be thus more effectively transported to the arctic regions. The further enhanced contaminant transport by currents is however uncertain since from the winter of 2010 NAO index tends to be low. The contaminant flux increase may be also caused by higher precipitation. The further global changes predictions show that the precipitation in the Arctic will increase (Macdonald et al. 2005). Thus, the "washing" of the contaminants from the atmosphere will be more effective (Schiedek et al. 2007).

2.2 Ocean Currents

A significant marine transport route for influxes of contaminants into the marine arctic environment is via inflow of water from the Norwegian Atlantic Current (NwAC) and the Norwegian Coastal Current (NCC). Transport with the NCC and the NwAC constitutes the dominant marine pathways for some contaminants originating from western Europe. The role of long-range oceanic transport may be significant for soluble, conservative radionuclides (e.g., Cs-137) (Gao et al. 2004; Johannessen et al. 2010), heavy metals (Cd, Pb) (Macdonald et al. 2005; Maccali et al. 2013) and organic chemicals with relatively low Henry's Law constants (e.g. HCH) (Lohmann et al. 2007; Halsall 2004).

While the NCC and the in-shore branch of the NwAC enter the western Barents Sea, the off-shore branch of the NwAC continues northwards. It flows into the Arctic Ocean through the Fram Strait. This branch sub-ducts under the fresh, cold layers of Polar Surface Water, separated by a strong vertical salinity gradient. The Barents Sea branch of the NwAC, mixed with the waters of the NCC, continues eastward across the Barents Sea Shelf to the Kara Sea. The warmer waters of Atlantic origin lose their heat and form dense waters that cascades down to the sea bottom. Thus contaminants that were transported by NwAC reach the bottom parts of the Eurasian Basin seawaters (Karcher et al. 2010).

The positive NAO index indicating increased in the strength Atlantic Oscillation may result in an enhanced northward delivery of contaminants from the Western Europe to the Arctic Ocean. The Norwegian Atlantic Current changes in volume and heat transfer are well documented (Hurrell and Deaer 2010). Higher transport of Atlantic origin water coupled with ice cover reduction will cause larger penetration of contaminated water masses into high latitudes. The effect of NAO may not be so certain since recently (from 2010) the NAO index is low.

2.3 Riverine Runoff

The mean annual riverine discharge to Arctic Ocean was estimated recently to 3,588 km^3 (Syed et al. 2007). The largest rivers discharging freshwater to the Arctic Ocean are Yenisei ($\sim$620 km^3 $year^{-1}$) and Lena ($\sim$530 km^3 $year^{-1}$) and Ob ($\sim$404 km^3 $year^{-1}$). However the suspended matter discharge to the Arctic Ocean is highest at Mackenzie (124 $\times$ 10^6 t $year^{-1}$), Yukon (54 $\times$ 10^6 t $year^{-1}$) and Lena (21 $\times$ 10^6 t $year^{-1}$) (Stein 2008). The freshwater run-off and the suspended matter influx to the Arctic Ocean is largest in May and June.

The Siberian rivers catchment areas include many diffuse contaminant sources such as agricultural runoff loaded with pesticides and discharges of municipal and industrial sewage from heavily populated and industrialized areas. Other polluting activities that pollute rivers are mining, gas exploitation (Harms et al. 2000), and radioactive contamination from the Mayak Production Association in the Urals, the Siberian Chemical Combine, and the Krasnoyarsk Mining and Chemical Combine (KMCC) (Melnikova et al. 2003; Lind et al. 2006; Skipperud et al. 2004, 2009). Rivers transport contaminants for long distances from southern parts of their catchment but there are also important point sources of contaminants located within Arctic. Particularly large mines and industrial processing plants in the Kola Peninsula, Pechora Basin, Urals, Noril'sk and Yakutsk introduce contaminated waters into the rivers system (Clarke and Harris 2003; Karcher et al. 2010).

The Arctic Ocean interior is characterized by a strong vertical haline stratification. Due to its low density, the major portion of the inflowing river water from the shelves stays in the upper tens of meters. Dissolved contaminants may reach deeper water layers only through vertical mixing or convective overturning in well-mixed shelf areas. The situation is different for particle bound contaminants (e.g. some organic contaminants, heavy metals and radionuclides). Up to 90 % of the suspended organic matter settle in the estuaries in contrast to 20–40 % of riverine dissolved organic components (Stein 2008).

Simulations have shown that annual river discharge will increase by 20 % for main arctic rivers (Arora and Boer 2001; Macdonald et al. 2005; Stein 2008). The increase of river runoff from riverine drainage basins into the Arctic Ocean may lead to increased inputs of contaminants from Siberian rivers. The peatlands covering Arctic and sub-Arctic areas over thousands of years have accumulated contaminants strongly associated with organic matter, such as mercury (Hg) (Biester et al. 2003; Bindler et al. 2004) and persistent organic pollutants, e.g., DDT and PCB (Turetsky et al. 2004). Most of peatlands are underlain by permafrost (Gorham 1991), constituting about 24 % of terrestrial surface in Northern Hemispere (Stein 2008). Due to actual temperature increase, permafrost have currently started thawing (Smith and Frey 2005; Fronzek et al. 2006; IPCC 2013; Turetsky et al. 2007). The rain water will now be able to penetrate soils containing contaminants and flush them into rivers. Higher riverine run-off coupled with intense permafrost melting will have increasing significance contaminants transport in the Arctic.

2.4 Sea Ice Transport

The ice gathers material from the atmosphere, by overflowing river water in spring or through freezing of bottom sediments (Pfirman et al. 1997; Harms et al. 2000). The general scenario is that the pollutants are entrained during the freezing process and encapsulated by the thickening ice. The ice cover of the Arctic Seas is contaminated by all types of contaminants associated to suspended matter transported with the river runoff (Borga et al. 2002; Masqué et al. 2007; Pavlov and Stanovoy 2001; Simoes and Zagorodnov 2001). After entertainment, the pollutants are transported by the motion of the sea ice, driven mainly by Transpolar Drift Stream. Ice exported from the Siberian Seas to the open Arctic Ocean is then transported through the Fram Strait towards the North Atlantic, where it melts (Masqué et al. 2007). The rate sea ice export is also influenced by NAO. The ice export is greater when NAO is positive (Rigor and Wallace 2004).

The Laptev Sea and Kara Sea as recognized as the main source of sediment-laden sea ice in the central Arctic (Pfirman et al. 1997; Rigor and Colony 1997; Hop and Pavlova 2008). Ice from the Kara and Laptev Seas may reach the Fram Strait in approximately 2–4 years, but sea ice entering the Beaufort Gyre may circulate even for about 5–15 years (Masqué et al. 2007).

The annual export of sediments to the North Atlantic by sea ice has been estimated to range from 7 to 150 × 10^6 tones. This sediment load together with associated contaminants is released during melting along marginal ice zones (Masqué et al. 2007). Contaminants released from sea ice may change regional distribution of contaminants in Fram Strait, the East Greenland shelf, or the Barents Sea (Pfirman et al. 1997; Zaborska et al. 2010).

The Arctic sea ice cover has been reduced substantially over the last decades both in terms of area, (with a decline in the summer sea ice extent of more than 10 % per decade between 1979 and 2007), and in terms of volume as indicated by an observed average thinning of more than 1.5 m since the 80s (Stranne and Björk 2012). The decrease of the ice cover will initially increase flux of contaminants incorporated into sea ice to the marine ecosystem. However, with the further ice cover and ice formation reduction this transport pathway for contaminants will have lower importance.

2.5 Coastal Erosion and Ground Water Discharge

There is little data concerning coastal erosion as discharge process for material to the marine environment. None of available papers reports discharge of contaminants related to shore erosion. The only studied regions are situated on the Barents Sea shore, these are Kolguev island, Cap Bolvansky (Vasiliev et al. 2005), and transect from Pesyakov Island to Meynski Zavrot Penisula (Leontyev 2003). Vasiliev et al. (2005) report that up to 50 m cliffs at the Kolguev island are built of fine material, sands and gravel. The shores are eroded with the speed of 0.4–3.4 m/year due strong action of waves. Lower (up to 10 m) marine terrace of Pesyakov Island and Meynski Zavrot Penisula studied by Leontyev (2003) retreats 1–4 m/year. He estimates that 5–15 m^3/m of the coast per year is lost due to erosion. It is possible that the coast soils inter-corporated some contaminants. Romankevich and Vetrov (2001) estimate that Barents Sea may receive 59×10^6 t $year^{-1}$ and Kara Sea may receive even 109×10^6 t $year^{-1}$ of material annually from coastal erosion.

The wave activity is expected to by higher in the Arctic Ocean due to less ice allowing longer fetch (Thomson and Rogers 2014). Increased wave activity will cause higher erosion and possible influx of contaminants to the sea, but it is really difficult to estimate their possible discharge to marine environment.

The degradation of coastal permafrost is believed to be important source of organic material (Grigoriev and Rachold 2003). Since organic contaminants are often associated with organic matter the thawing of coastal permafrost may become the important source of these contaminants. Enhanced permafrost thawing may also increase ground water discharge. Submarine groundwater discharge (SGD) has been recognized as an important factor influencing coastal zones and significant source of contaminants to the marine environment (Gallagher et al. 1996; Szymczycha et al. 2013).

2.6 Biological Transport Driven by Migrating Animals

Persistent Organic Pollutants, some trace elements (e.g. Hg) and radionuclides are assimilated into lower level producers (phytoplankton) and then biomagnified up trophic chains, making higher tropic level species more vulnerable to contaminant exposure via their diet. Many Arctic animals disperse or migrate and thus are exposed to different levels and types of contaminants (Brunström and Halldin 2000). Migrating animals are transporting man-made substances from middle latitudes to the Arctic. This process concerns mainly birds, fish and marine mammals.

The highest trophic level animals e.g. seals, whales, polar bears accumulate and may transport contaminants, however such transport have rather local significance. However birds can transfer contaminants fro very large distances. For instance glaucous gull and fulmars have circumpolar distribution and may transfer contaminants in their tissues (Brunström and Halldin 2000). Birds feeding at the sea can not only transfer contaminants for long distances but also transfer toxins from marine to terrestrial environment (Blais et al. 2005; Evenset et al. 2004).

3 The Influence of Environmental Properties Variability on Contaminants Distribution in Marine Arctic Ecosystem

3.1 Temperature Raise

Increased temperature will have direct effect on contaminants cycling. The most important direct effect may be more rapid degradation of organic contaminants (Macdonald et al. 2005). Contaminant degradation processes are less efficient in the Arctic due to low temperature and low solar radiation input. That is why most of contaminants (POPs) persist in the Arctic for longer time than in warm and temperate regions (Bard 1999). More efficient degradation would be generally positive process but is should be remembered that some degradation products may be more toxic than their parent substances (Coats 1993). The good example is DDT that was used as pesticide and it's degradation product—DDE.

Positive consequence of increased temperature may be the alterations in species energetics and trophic interactions. It was previously reported that adaptations to the arctic climate imply an increased lipid production. Arctic marine organisms have developed specific life cycles to match the productive period and the ability to store energy in lipids (Falk-Petersen et al. 1987; Borga et al. 2002). Thus, the transfer of energy between arctic trophic levels is more effective compared to temperate regions. With the temperature rise the organisms will not produce so much lipids and the change of energy transfer between trophic levels is expected (Sobek et al. 2010). Since POPs accumulate in organisms in lipids, their transfer in the trophic chain may be limited. Other advantage of temperature raise may be dilution of organic pollutants in biological tissues due to growing body mass so called "growth dilution" (Gobas 1993; Sobek et al. 2010).

3.2 Enhanced UV Radiation

Due to short periods of sun light presence arctic organisms have adapted to low UV-light levels and show higher photosynthetic efficiencies at lower UV-light levels and may grow even under ice cover. Organisms living in association with the sea ice may be exposed to contaminants accumulated in sea ice, especially when it melts (Pfirman et al. 1990; Weeks 1994). Thus, ice associated organisms may be exposed to larger levels of contaminants during summer comparing to temperate organisms.

With the ice cover decrease, the UV light and sunlight penetration into water column will be larger. This will enhance the photodegradation of organic contaminants (Dąbrowska et al. 2004; Brillas 2014), and will help to eliminate a part of POPs more vulnerable to degradation. It was also found that degradation of extremely toxic form of mercury—methylmercury will accelerate with a loss of the arctic ice cover by a larger penetration of sunlight (Point et al. 2011).

3.3 Oxygen Concentration, pH and Redox Conditions

Changes in pH and redox conditions strongly influence chemical forms of radionuclides (e.g. Pu, Cs) and heavy metals (e.g. Hg). Their different forms—oxidation states—influence distinct affinity to suspended particulate matter. For instance, in reducing environment (low pH, low redox, low oxygen) plutonium exists at IV of V oxidation state (Mitchell et al. 1991; Choppin and Morgenstern 2001). Reduced plutonium is highly particle reactive and is readily scavenged from the water column to sediments. In oxidizing environment, the Pu is exist in form of PuO_2^+ and is very soluble in the water column. In that form it may be transported for the long distances and its elimination to marine sediments is very limited.

Usually oxygen is present only in surface most sediment layers and contaminants deposited in deeper layers many years ago are immobile. However, mixing processes caused by animals (bioturbation) or physical processes (bottom currents), allow the oxygen enter to the lower sediment layers. Sediment mixing may facilitate re-mobilization of contaminants and their re-introduction to sediment pore waters and then water column (Thibodeaux and Bierman 2003) and may induce changes of chemical form from less to more bioavailable.

The possible effects of global changes: the increase of bottom fauna population and the higher intensity of physical processes (storms) may facilitate contaminant re-introduction to marine environment. Thus, bottom sediments that have been a sink for contaminants for the last century may now become their sources (Vintró et al. 2002).

3.4 Change in Salinity

The salinity changes in the Arctic depend on the global change scenario, but the future trends of salinity is poorly constrained. With the higher influence of Atlantic waters the warmer, more saline waters will penetrate larger Arctic areas. However, the actually observed temperature rise initiates glaciers and ice cover melting, thus the arctic marine environment will receive the enormous influx of freshwater.

Most of the studies indicate that toxicity of inorganic contaminants increases with decreasing salinity (Hall and Anderson 1995; Ytreberg et al. 2011). Heavy metals (e.g. Cd, Hg, Zn) are more bioavailable in lower salinity due to presence of free metal ions. Increasing salinity decreases the dissolved concentrations of metal ions due to the precipitation of the metal ions. Thus in consequence in more saline waters the toxicity of metals to organisms is reduced (Park et al. 2014).

Some organic contaminants e.g. organophosphate insecticides, showed to be more toxic with increased salinity. It was also found that fish were more resistant to toxic chemicals at middle salinities when compared with low salinities (Hall and Anderson 1995). However, Ramachandran et al. (2006) indicate much higher toxicity of PAH in less saline as compere to more saline waters. This phenomena is caused by increased solubility of PAH at lower salinities, especially the lower molecular weight two- and three-ringed PAH homologs. This solubility effect is enhanced by the apparent increased of effectiveness of chemical dispersion at low salinities.

4 Increased Human Activity in the Arctic

4.1 Enhanced Shipping and Tourism

The decrease of ice cover will encourage shipping and tourism. The Arctic is favored shipping route between Europe, North America and Asia (Macdonald et al. 2005). Routine shipping is connected to usage of oil, the source of hydrocarbons. The enhanced transport activities create also risks of maritime accidents and accidental fuel leaks.

4.2 Oil Installations

The recent and further decrease of ice cover will allow industrial activities to begin in the Arctic. It has been estimated that $100 billion could be invested in the Arctic over the next decade. The intense discussion on the oil exploration in the Arctic Ocean have been already started. The Arctic contains vast oil and gas constituting 13 and 30 % of the world's undiscovered resources respectively (Eurasia group report 2013). The incidents of oil spill connected to oil installations operation and oil shipping happen every year in different marine areas. The report published by International Tanker Owners Pollution Federation (ITOPF Report 2014), indicates that the annual average number of spills greater than 700 tonnes in the all oceans was 3–4 per year during 2000–2009 and 1–2 spills per year during 2010–2012. Part of the oil spill undergoes evaporation and dispersion in the seawater leading to oil removal or dilution. Unfortunately, part of an oil spill undergoes emulsification and sedimentation and in persist forms enhances local marine environment contamination (ITOPF Report 2014).

Some attention should be given to the produced water from oil and gas installations that contains high concentrations of radium isotopes. These isotopes are naturally occurring elements in geological formations containing gas and oil. The contaminated water produced during oil extraction is discharged to the sea and radium and its decay products may be available for the marine organisms (Eriksen et al. 2009).

4.3 Nuclear Power Installations

The other danger for the Arctic environment is connected to potential accidents in nuclear power plants and installations, decommissioning and refueling of nuclear power vessels. Until now several accidents has happened with nuclear powered submarines e.g. Kursk however their impact to the environment was negligible (Amundsen et al. 2002; Matishov et al. 2002). The possible accident with a nuclear fuel leakage to marine environment is possible and in consequence the radionuclides levels in organisms near the accident location may exceed recommended levels (Iosjpe et al. 2011).

The Russian Federation had recently increased the number of nuclear powered vessels and installation in the Arctic. Particularly, the development of floating nuclear power plants (FNPPs) off-shore Siberia is believed to pose a potential danger to the environment (Kuznetsov et al. 2004). The risk of leakage of radioactive materials, explosion or terrorist attack would be greatly increased when the FNPPs system project is completed (one FNPP "Akademik Lomonosov" is already launched, several are planned to be finished by the end of this decade).

4.4 Re-introduction of Deposited Contaminants

Several activities performed by humans may disrupt the marine sediments and induce the re-mobilisation of contaminants archived in deep sediment layers. The increase of bottom fishery using trawling techniques may disrupt large areas the Arctic Sea bottom. The deep mixing of shelf sediments may be also induced by underwater constructions e.g. oil exploitation platforms, submarines or floating nuclear power plants. The contaminants re-introduction may be also caused by biological activity (bioturbation) (Thibodeaux and Bierman 2003). Thus, the increase of benthic fauna biomass caused by less demanding environmental conditions may increase the chemicals release.

5 Summary

Global changes such as temperature rise, air and water circulation patterns or precipitation patterns changes, will have direct and indirect influence on contaminants distribution in the Arctic. Contaminants long range transport, emissions and sinks, persistence and degradation processes and speciation will be definitely modified. The fate of contaminants, their accumulation in organisms and toxicology will undergo changes together with marine food web structure or animal migration habits. The large risk of new contamination of the Arctic environment is also connected to increase of human activity in the Arctic.

References

Abram NJ, Wolff EW, Curran MAJ (2013) A review of sea ice proxy information from polar ice cores. Quat Sci Rev 79:168–183

Amundsen I, Iosjpe M, Reistad O et al (2002) The accidental sinking of the nuclear submarine, the Kursk: monitoring of radioactivity and the preliminary assessment of the potential impact of radioactive releases. Mar Pollut Bull 44(6):459–468

Arora VK, Boer GJ (2001) Effects of simulated climate change on the hydrology of major river basins. J Geophys Res. doi:10.1029/2000JD900620

Bard SM (1999) Global transport of anthropogenic contaminants and the consequences for the Arctic marine ecosystem. Mar Pollut Bull 38(5):356–379

Biester H, Cortizas MA, Birkenstock S et al (2003) Effect of peat decomposition and mass loss on historic mercury records in peat bogs from Patagonia. Environ Sci Technol 37:32–39

Bindler R, Klarqvist M, Klaminder J et al (2004) Does within-bog spatial variability of mercury and lead constrain reconstructions of absolute deposition rates from single peat records? The example of Store Mosse, Sweden. Glob Biogeochem Cycles 18(3). doi:10.1029/2004GB002270

Blais JM, Kimpe LE, McMahon D et al (2005) Arctic seabirds transport marine-derived contaminants. Science 309:445

Borga K, Poltermann M, Polder A et al (2002) Influence of diet and sea ice drift on organochlorine bioaccumulation in Arctic ice-associated amphipods. Environ Pollut 117:47–60

Brillas E (2014) A review on the degradation of organic pollutants in waters by UV photoelectro-fenton and solar photoelectro-fenton. J Braz Chem Soc 25(3):393–417

Brunström B, Halldin K (2000) Ecotoxicological risk assessment of environmental pollutants in the Arctic. Toxicol Lett 112–113:111–118

Choppin GR, Morgenstern A (2001) Distribution and movement of environmental plutonium. Radioact Environ 1:91–105

Clarke A, Harris CM (2003) Polar marine ecosystems: major threats and future change. Environ Conserv 30(1):1–25

Coats JR (1993) What happens to degradable pesticides? ChemTech 5:25–29

Crane K, Galasso J, Brown C et al (2000) Northern ocean inventories of radionuclide contamination: GIS efforts to determine the past and present state of the environment in and adjacent to the Arctic. Mar Pollut Bull 40(10):853–868

Dąbrowska D, Kot-Wasik A, Namieśnik J (2004) The importance of degradation in the fate of selected organic compounds in the environment. Part II. Photodegradation and biodegradation. Pol J Environ Stud 13(6):617–626

Dowdall M, Vicat K, Frearson I, Gerland S, Lind B, Shaw G (2004) Assessment of the radiological impacts of historical coal mining operations on the environment of Ny-Alesund, Svalbard. J Environ Radioact 71(2):101–114

Elberling B, Asmund G, Kunzendorf H, Krogstad EJ (2002) Geochemical trends in metal-contaminated fiord sediments near a former lead–zinc mine in West Greenland. Appl Geochem 17:493–502

Efurd DW, Steiner RE, Roensch FR et al (2005) Determination of the $^{240}Pu/^{239}Pu$ atom ratio in global fallout at two locations in the Northern Hemisphere. J Radioanal Nucl Chem 263 (2):387–391

Eurasia Group Report (2013) Opportunities and challenges for Arctic oil and gas development. Report for Wilson Center, Washington DC. http://www.wilsoncenter.org/sites/default/files/Artic%20Report_F2.pdf

Eriksen DØ, Sidhu R, Ramsøy T et al (2009) Radioactivity in produced water from Norwegian oil and gas installations—concentrations, bioavailability, and doses to marine biota. Radioprotection 44(5):869–874

Evenset A, Christensen GN, Skotvold T et al (2004) A comparison of organic contaminants in two high Arctic lake ecosystems, Bjbrnbya (Bear Island), Norway. Sci Total Environ 318:125–141

Falk-Petersen S, Sargent JR, Tande K (1987) Food pathways and life strategy in relation to the lipid composition of sub-Arctic zooplankton. Polar Biol 8:115–120

Fronzek S, Luoto M, Carter TR (2006) Potential effect of climate change on the distribution of palsa mires in subarctic Fennoscandia. Clim Res 32:1–12

Gallagher DI, Dietrich AM, Reay WG et al (1996) Ground water discharge of agricultural pesticides and nutrients to estuarine surface water. Ground Water Monit Rem 16:118–129

Gao Y, Drange H, Bentsen M, Johannessen OM (2004) Simulating transport of non-chernobyl 137Cs and 90Sr in the North Atlantic–Arctic region. J Environ Radioact 71:1–16

Gobas FAPC (1993) A model for predicting the bioaccumulation hydrophobic organic chemicals in aquatic food-webs: application to lake Ontario. Ecol Model 69:1–17

Gobeil C, Macdonald RW, Smith JN et al (2001) Lead contamination in Arctic basin sediments tracks Atlantic water flow pathways. Science 293:1301–1304

Gordeev VV (2002) Pollution of the Arctic. Reg Environ Change 3:88–98. doi:10.1007/s10113-002-0041-4

Gorham E (1991) The role of northern peatlands in the carbon cycle, and their probable response to climate warming. Ecol Appl 1:182–195

Grigoriev MN, Rachold V (2003) The degradation of coastal permafrost and the organic carbon balance of the Laptev and East Siberian Seas. In: Phillips M, Springman SM, Arenson LU (eds) Permafrost. Swets & Zeitlinger, Lisse

Hall LW, Anderson RD (1995) The influence of salinity on the toxicity of various classes of chemicals to aquatic biota. Crit Rev Toxicol 25(4):281–346

Halsall CJ (2004) Investigating the occurrence of persistent organic pollutants (POPs) in the arctic: their atmospheric behaviour and interaction with the seasonal snow pack. Environ Pollut 128 (1–2):163–175. doi:10.1016/j.envpol.2003.08.026

Halsall CJ, Sweetman AJ, Barrie LA, Jones KC (2001) Modelling the behaviour of PAHs during atmospheric transport from the UK to the Arctic. Atmos Environ 35:255–267

Halsall CJ, Bailey R, Stern GA et al (1998) Multi-year observations of organohalogen pesticides in the Arctic atmosphere. Environ Pollut 102:51–62

Harms IH, Karcher MJ, Dethleff D (2000) Modelling Siberian river runoff—implications for contaminant transport in the Arctic Ocean. J Mar Syst 27:95–115

Hurrell JW, Deaer C (2010) North Atlantic climate variability: the role of the north Atlantic oscillation. J Mar Syst 79:231–244

Hop H, Pavlova O (2008) Distribution and biomass transport of ice-amphipods in Svalbard waters. Deep-Sea Res II 55(20–21):2292–2307. doi:10.1016/j.dsr2.2008.05.023

Iosjpe M, Reistad O, Liland A (2011) Radioecological consequences after a hypothetical accident with release into the marine environment involving a Russian nuclear submarine in the Barents Sea. Strålevern Rapport 2011:3

ITOPF (2014) Fate of marine oil spills. Technical information paper. http://www.itopf.com/knowledge-resources/documents-guides/document/tip-2-fate-of-marine-oil-spills/

IPCC 2013: Climate Change 2013: the physical science basis. Contribution of working group I to the fifth assessment report of the intergovernmental panel on climate change. In: Stocker TF, Qin D, Plattner G-K, Tignor M, Allen SK, Boschung J, Nauels A, Xia Y, Bex V, Midgley PM (eds) Cambridge University Press, Cambridge, UK and New York, NY, USA, 1535 pp

Johannessen OM, Bengtsson L, Miles MW et al (2004) Arctic climate change: observed and modeled temperature and sea-ice variability. Tellus 56A:328–341

Johannessen OM, Volkov VA, Pettersson LH, Maderich VS, Zheleznyak MJ, Gao Y, Bobylev LP, Stepanov AV, Neelov IA, Tishkov VP and Nielsen SP (2010) Radioactivity and pollution in the Nordic seas and Arctic Region. observations, modeling and simulation. Springer, Nansen Center's Polar Series, p 408

Karcher M, Harms I, Standring WJF (2010) On the potential for climate change impacts on marine anthropogenic radioactivity in the Arctic regions. Mar Pollut Bull 60:1151–1159

Kuznetsov VM, Yablokov AV, Kolton IB et al (2004) Floating nuclear power plants in Russia: a threat to the Arctic, world oceans and non-proliferation treaty. Agenstwo Rakurs Production Moscow. http://www.greencross.ch/uploads/media/gc_fnpp_book.pdf

Leontyev IO (2003) Modeling erosion of sedimentary coasts in the western Russian Arctic. Coast Eng 47:413–429

Lind OC, Oughton DH, Salbu B et al (2006) Transport of low Pu-240/Pu-239 atom ratio plutonium-species in the Ob and Yenisey Rivers to the Kara Sea. Earth Planet Sci Lett 251:33–43

Lohmann R, Breivik K, Dachs J et al (2007) Global fate of POPs: current and future research directions. Environ Pollut 150:150–165

Maccali J, Hillaire-Marcel C, Carignan J et al (2013) Geochemical signatures of sediments documenting Arctic sea-ice and water mass export through Fram Strait since the last glacial maximum. Quatern Sci Rev 64:136–151

Macdonald RW, Harner TT, Fyfe J (2005) Recent climate change in the Arctic and its impact on contaminants pathway and interpretation on temporal trend data. Sci Total Environ 342:5–86

Masqué P, Cochranb JK, Hirschberg DJ (2007) Radionuclides in Arctic sea ice: tracers of sources, fates and ice transit time scales. Deep-Sea Res I 54:1289–1310

Matishov GG, Matishov DG, Namiatov AE et al (2002) Radioactivity near the sunken submarine "Kursk" in the southern Barents. Sea Environ Sci Technol 36(9):1919–1922

McBean G, Alekseev G, Chen D (2004) Arctic climate: past and present, ACIA Chapter 2. http://www.acia.uaf.edu/PDFs/ACIA_Science_Chapters_Final/ACIA_Ch02_Final.pdf

Melnikova S, Carrollb JJ, Gorshkov A (2003) Snow and ice concentrations of selected persistent pollutants in the Ob–Yenisey River watershed. Sci Total Environ 306:27–37

Mitchell PI, Battle JV, Ryan TP et al (1991) Studies on the speciation of plutonium and americium in the western Irish Sea. In: Kershaw PJ, Woodhead DS (eds) Radionuclides in the study of marine processes. Elsevier Applied Science, London

Park J, Kim S, Yoo J et al (2014) Effect of salinity on acute copper and zinc toxicity to Tigriopus japonicus: the difference between metal ions and nanoparticles. Mar Pollut Bull 85:526–531

Pavlov VK, Stanovoy VV (2001) The problem of Transfer of radionuclide pollution be sea ice. Mar Pollut Bull 42(4):319–323

Pfirman S, Lange MA, Wollenburg I et al (1990) Sea ice characteristics and the role of sediment inclusions in deepsea deposition: Arctic-Antarctic comparisons. Geological history of the polar oceans: Arctic versus Antarctic. Kluwer Academic Publishers, Dordrecht, pp 187–211

Pfirman SL, Kögeler JW, Rigor I (1997) Potential for rapid transport of contaminants from the Kara Sea. Sci Total Environ 202(1–3):111–122

Point D, Sonke JE, Day RD et al (2011) Methylmercury photodegradation influenced by sea-ice cover in Arctic marine ecosystems. Nat Geosci 4:188–194

Pogrebov VB, Fokin SI, Galtsova VV et al (1997) Benthic communities as influenced by nuclear testing and radioactive waste disposal off Novaya Zemlya in the Russian Arctic. Mar Pollut Bull 35:333–339

Ramachandran SD, Sweezey MJ, Hodson PV et al (2006) Influence of salinity and fish species on PAH uptake from dispersed crude oil. Mar Pollut Bull 52(10):1182–1189

Rigor I, Colony R (1997) Sea-ice production and transport of pollutants in the Laptev Sea, 1979-1993. Sci Total Environ 202:89–110

Rigor I, Wallace JM (2004) Variations in the age of Arctic sea-ice and summer sea-ice extent. Geophys Res Lett 31:L09401

Romankevich EA, Vetrov AA (2001) Cycle of carbon in the Russian Arctic Seas. Nauka, Moscow, p 302 (in Russian)

Rydberg J, Klaminder J, Rosén P et al (2010) Climate driven release of carbon and mercury from permafrost mires increases mercury loading to sub-arctic lakes. Sci Total Environ 408:4778–4783

Schiedek D, Sundelin B, Readman JW et al (2007) Interactions between climate change and contaminants. Mar Pollut Bull 54:1845–1856

Simoes JS, Zagorodnov VS (2001) The record of anthropogenic pollution in snow and ice in Svalbard, Norway. Atmos Environ 35:403–413

Skipperud L, Brown J, Fifield L et al (2009) Association of plutonium with sediments from the Ob and Yenisey Rivers and Estuaries. J Environ Radioact 4:290–300

Skipperud L, Oughton DH, Fifield LK et al (2004) Plutonium isotope ratios in the Yenisey and Ob estuaries. Appl Radiat Isot 60(2–4):589–593
Smith JN, Ellis KM, Kilius LR (1998) ^{129}I and ^{137}Cs tracer measurements in the Arctic Ocean. Deep-Sea Res I 45:959–984
Smith JN, Ellis KM, Polyak L et al (2000) ^{239,240}Pu transport into the Arctic Ocean from underwater nuclear tests in Chernaza Baz, Novaza Yemlza. Cont Shelf Res 20:255–279
Smith LC, Frey KE (2005) Amplified carbon release from vast West Siberian peatlands by 2100. Geophys Res Lett 32. doi:10.1029/2004GL022025
Sobek A, McLachlan MS, Borgå K et al (2010) A comparison of PCB bioaccumulation factors between an arctic and a temperate marine food web. Sci Total Environ 408:2753–2760
Stein R (2008) Arctic ocean sediments: processes, proxies, and paleo environment. developments in Marine Geology, vol 2. Elsevier, Amsterdam, p 592
Stranne C, Björk G (2012) On the Arctic Ocean ice thickness response to changes in the external forcing. Clim Dyn 39:3007–3018. doi:10.1007/s00382-011-1275-y
Syed TH, Famiglietti JS, Zlotnicki V, Rodell M (2007) Contemporary estimates of Pan-Arctic freshwater discharge from GRACE and reanalysis. Geophys Res Lett 34:L19404. doi:10.1029/2007GL031254
Szymczycha B, Miotk M, Pempkowiak J (2013) Submarine groundwater discharge as a source of mercury in the Bay of Puck. Water Soil Air Pollut 224:1542. doi:10.1007/s11270-013-1542-0
Thibodeaux LJ, Bierman VI (2003) The bioturbation-driven chemical release process. Environ Sci Technol 37(13):252–258. doi:10.1021/es032518j
Thomson J, Rogers WE (2014) Swell and sea in the emerging Arctic Ocean. Geophys Res Lett 41:3136–3140. doi:10.1002/2014GL059983
Turetsky MR, Manning SW, Wieder RK (2004) Dating recent peat deposits. Wetlands 24:324–356
Turetsky MR, Wieder RK, Vitt DH et al (2007) The disappearance of relict permafrost in boreal north America: effects on peatland carbon storage and fluxes. Glob Change Biol 13:1922–1934
Vasiliev A, Kanevskiy M, Cherkashov G et al (2005) Coastal dynamics and the Barents and Kara Sea key sites. Geo-Mar Lett 25(2–3):110–120
Vavrus S, Harrison SP (2003) The impact of sea-ice dynamics on the Arctic climate system. Clim Dyn 20:741–757. doi:10.1007/s00382-003-0309-5
Vintró LL, McMahon CA, Mitchell PI et al (2002) Transport of plutonium in surface and subsurface waters from the Arctic shelf to the North Pole via the Lomonosov Ridge. J Environ Radioact 60(1–2):73–89
Weeks W (1994) Possible roles of sea ice in the transport of hazardous material. Interagency Arctic Res Policy Committee 8:34–52
Wania F, Su Y (2004) Quantifying the global fractionation of polychlorinated biphenyls. Ambio 33(3):161–168
Wit CA, Muir D (2010) Levels and trends of new contaminants, temporal trends of legacy contaminants and effects of contaminants in the Arctic: preface. Sci Total Environ 408:2852–2853
Ytreberg E, Karlsson J, Ndungu K et al (2011) Influence of salinity and organic matter on the toxicity of Cu to a brackish water and marine clone of the red macroalga ceramium tenuicorne. Ecotoxicol Environ Saf 74:636–642
Zaborska A, Mietelski JW, Carroll J et al (2010) Sources and distributions of ^{137}Cs, ^{238}Pu, ^{239,240}Pu radionuclides in the north-western Barents Sea. J Environ Radioact 101:323–331
Zaborska A, Carroll J, Pazdro K et al (2011) Spatio-temporal patterns of PAHs, PCBs and HCB in sediments of the western Barents Sea. Oceanologia 53(4):1005–1026

The Adaptations of the Foraminifera and Ostracoda to Fresh Water Colonisation

Anna Iglikowska and Joanna Pawłowska

Abstract In marine environments Ostracoda and Foraminifera have been very successful invaders. During the Phanerozoic they colonised the majority of shallow, marginal to deep water, fully marine habitats. Both groups had developed physiological adaptations which pre-adapted them to the invasion of new marine habitats. They adopted a broad range of feeding strategies and reproduction modes. The production of resting stages and brood care may also have contributed to them being efficient invaders. They are also both highly tolerant to variations in salinity. The first invasions of non-marine habitats by ostracods appear to have taken place at the turn of the Devonian and Carboniferous. It is estimated that there had been between 9 and 12 independent invasions of fresh waters by the ostracods. In contrast Foraminifera are typically marine organisms, and only a few species of agglutinated and organic-walled Foraminifera are to be found in brackish and freshwater environments. Agglutinated species build their test using ambient components but are not commonly regarded as calcifying organisms. An impact of salinity on foraminiferal calcification has been observed in several studies. It seems that Foraminifera are incapable of constructing a fully calcified test in low salinity regimes; they use sea water not only as a source of ions to construct shell, but also as a biomineralisation solution. Thus, the success of ostracods in invading freshwater habitats can be attributed to their development of a more effective mechanism of calcification in low mineralisation waters. The core question of this study is to examine possible causes for the differences in success between the two taxa.

Keywords Foraminifera · Ostracoda · Fresh water colonisation · Physiological adaptations · Calcification

---

A. Iglikowska (✉) · J. Pawłowska
Institute of Oceanology of Polish Academy of Sciences, Powstańców Warszawy 55, 81-712 Sopot, Poland
e-mail: iglikowska@iopan.gda.pl

J. Pawłowska
e-mail: pawlowska@iopan.gda.pl

T. Zielinski et al. (eds.), *Impact of Climate Changes on Marine Environments*,
GeoPlanet: Earth and Planetary Sciences, DOI 10.1007/978-3-319-14283-8_8

1 Introduction

This is a review paper focused on the invasion of non-marine habitats by Foraminifera and Ostracoda. The paper reviews the comprehensive paleontological and modern data and integrates these with the knowledge of physiological adaptations, which have enabled them to colonise of freshwater habitats. Both these groups have undergone spectacular adaptive radiations and have invaded a wide variety of marine habitats. In the Early Carboniferous ostracods and Foraminifera thrived, and both groups faced an opportunity to colonise freshwater habitats. While the Ostracoda succeeded, the Foraminifera have remained an almost exclusively marine group. The aim of this study was to elucidate what could be the crucial ability that limited the Foraminifera from successfully colonising freshwater habitats, and why ostracods were more efficient in the invasion of low salinity environments.

The earliest fossil Foraminifera appeared in the Early Cambrian, but molecular data indicate a much earlier, Neoproterozoic, origin (Pawlowski et al. 2003). The origin of the Ostracoda may be slightly younger. Evidence from molecular studies suggests that ostracods diverged from near the base of the Pancrustacea during the late Proterozoic (Regier et al. 2005; Siveter 2008; Williams et al. 2008), although the oldest unequivocal fossil record is from the mid-late Cambrian (Harvey et al. 2012).

The Foraminifera seems to be an enormously successful invader of new environments. During the Phanerozoic they colonised most shallow, marginal to deep water, fully marine habitats, and diversified to exploit a wide variety of life modes (Hottinger 1982; Goldstein 2003). Some attained relatively gigantic size, such as the extinct *Lepidocyclina elephantina* at 14 cm (Grell 1973). The mean size of modern Foraminifera ranges from 0.1 to 0.5 mm, however, some species may reach up to several centimetres (Pawlowski 2009). Despite their unicellular level of organization foraminiferans perform the same range of fundamental functions as metazoans (Goldstein 2003). There are two characteristic features that help to distinguish the Foraminifera from other protists. First, all possess granuloreticulopodia (pseudopodia) which are used for motion, feeding, constructing a test, protection and for some aspects of reproduction. Second, almost all Foraminifera have a test which encases the body, separating it from the surrounding environment. There are three different types of tests: organic, agglutinated (constructed from cemented particles) and mineralised, composed of calcium carbonate or, in rare cases, of silica (Goldstein 2003; Pawlowski et al. 2013).

Ostracods are also efficient colonisers of new habitats. The wide geographical distribution and their almost simultaneous appearance on several palaeocontinents suggest rapid dispersal and wide environmental tolerance (Williams et al. 2008). By the mid Silurian originally benthic myodocopes had started colonising the pelagic (Siveter et al. 1991; Vannier and Abe 1992; Perrier et al. 2011) and by the turn of the Devonian and Carboniferous they had invaded inland freshwater habitats and the deep ocean (Williams et al. 2006; Bennet 2008). Martens et al. (2008) estimated that ostracods had undergone between 9 and 12 independent incursions into fresh water from the marine environment.

As crustacean metazoans, ostracods have specialised tissues, and organ systems. They are typically larger than recent foraminiferans, mostly in the 0.3–5 mm range, although some marine species exceed 30 mm in length. The most distinctive feature of ostracods is their calcified carapace comprising paired, dorsally articulated valves (Meisch 2000). Both groups, Ostracoda and Foraminifera, have a marine origin, have a test or shell saturated with calcium carbonate, and inhabit aquatic environments. Both have undergone spectacular adaptive radiation and colonised wide variety of marine habitats. In the Early Carboniferous ostracods and foraminiferans thrived. It is likely that during peak marine transgression both groups had opportunities to colonise freshwater, inland habitats. Ostracods succeeded and foraminiferans seemed to be less fortunate, but why?

2 Evolution from the Palaeozoic to the Recent

Traditionally, the evolution of Foraminifera is viewed as a gradual process of change in structure and composition of the test, starting from naked, unilocular forms, via organic-walled and agglutinated forms which later became multilocular, and ending with the highly complex calcareous forms (Tappan and Loeblich 1988). Recent molecular studies (Pawlowski and Holzmann 2002; Pawlowski et al. 2003), however, revealed that there is no evidence for a progressive increase in foraminiferal test complexity. In a single highly supported clade, for example, a distinctive radiation included a wide variety of test morphotypes containing both agglutinated (Textulariida) and calcareous (Rotaliida) species. In a recent study Pawlowski et al. (2013) showed that transition from organic to agglutinated walls occurred several times, and the change in the nature of test wall was dependent on environmental conditions. According to these authors a calcareous wall appeared at least five times independently, and each time a different type of calcareous test was developed. It appears that some species may have lost their calcified test secondarily, for instance, as an adaptation to the lower salinity regime.

The first fossils of Foraminifera are agglutinated Textulariina and appeared during the Cambrian (McIlroy et al. 2001; Boudagher-Fadel 2008). They remained as a dominant group until the Silurian when the larger, calcareous and more complex Fusulinina appeared, becoming abundant in the late Palaeozoic. In the Silurian, apart from textulariines and fusulinines, a new group of Foraminifera with the test wall consisting of calcite crystals evolved—the Lagenida. Less significant ecologically in the Palaeozoic were the Miliolida (having porcelaneous test) and early Involutinina (with an aragonitic wall). More advanced calcareous tests, formed by biomineralisation of an inner tectinous lining, may have appeared by the Silurian, but did not become widespread until the late Devonian (McIlroy et al. 2001).

The Palaeozoic seems to be a period of fusulinines. They underwent rapid evolutionary radiation from tiny, simple organisms to large, complex and highly specialised forms in diverse lineages (eight known families) (Boudagher-Fadel 2008). Fusilinines became ubiquitous and spread to most warm, shallow waters from the Mississippian (≈325 Myr, Carboniferous) to the end of the Permian, when they may have become extinct (Payne et al. 2012), although, there is an alternative hypothesis (Leven 2010) that the primitive fusulinids, which gave rise to the superorder Fusulinoida, did not go extinct, but survived through transformation into more progressive taxa.

The success of Fusulinida in the Palaeozoic most likely was a result of their acquisition of a calcareous test. This test was primitively homogeneously microgranular and consisted of low-magnesium calcite. In advanced forms the test was larger (up to 15 cm) and had two or more differentiated layers. Fusulinines are among the largest single-celled protists preserved in fossil deposits, and provide perhaps the best-known case of evolution towards large size (Newell 1949; Payne et al. 2009, 2012). They are usually recorded in limestone sediments (Gallagher 1998; Leven and Gorgij 2011), and thus, probably had easy access to calcium-rich sea water as a resource for test construction.

In the Palaeozoic a variety of other ostracod-like groups of arthropods co-occurred with ostracods. Bradoriida, Phosphatocopida and Leperditicopida, are all characterised by having a calcified, bivalved carapace, but their systematic position remains unclear. The oldest carapaces of ostracods (Palaeocopida) are found in the early Ordovician (Williams et al. 2008). Early myodocopes appeared in the Silurian, but because of their poorly mineralised valves the early fossil record of this group is sparse (Siveter 2008). Podocopes were common in the Palaeozoic, with hundreds of species recorded, even in the Ordovician (Siveter and Curry 1984; Siveter 2008). Most of them are known from carapaces only, but rare examples of exceptionally preserved ostracods with soft body parts are known from the Mesozoic and younger deposits (Smith 2000; Siveter 2008). It is notable that the first non-marine ostracods were Podocopes. The systematic position of Palaecopida is still under debate, but they were a common and widespread ostracod-like group in the Palaeozoic (Gray 1988). Siveter (2008) claims that soft part anatomy of *Nymphatelina gravida* is similar to that of myodocopes, and it is possible that Palaeocopida may be an artificial group. Currently their distinctive valve morphology distinguishes them from other ostracod taxa (Siveter 2008; Siveter et al. 2010).

Marine invertebrates invaded non-marine environments multiple times since the Cambrian. Most probably the first major colonisation occurred during the Devonian-Carboniferous transition. The first brackish/freshwater habitats were near-shore, marine embayments or shallow, deltaic lagoons influenced by marine transgression events (Tibert and Scott 1999; Bennett et al. 2012). The first non-marine invaders were probably highly tolerant euryhaline species, capable of thriving in marginal marine environments with varying salinity regimes.

Tibert and Scott (1999) documented the early Carboniferous ostracods and foraminiferans of Horton Bluff Formation in Maritimes Basin (Nova Scotia, Atlantic Canada). This environment was interpreted as a restricted marine embayment that turned brackish. The marginal marine bay fauna was dominated by an assemblage of euryhaline marine ostracod species *Copelandella novascotica*, *Cavellina* sp., *Geisina* sp. and opportunistic paraparchitacean ostracods (*Shemonaella scotoburdigalensis*, *S. tatei* and *Chamishaella* sp.). The coastal pond was inhabited mainly by freshwater species such as *Carbonita scalpellus* and *C. rankiniana*. Foraminifera were recorded only in the coastal marsh, and they were represented exclusively by agglutinated species of *Trochammina* sp., as the dominant species, and by *Ammobaculites* sp., *Ammotium* sp. and *Ammodiscus* sp. which were less abundant.

A similar ecological system in Nova Scotia was studied by Calder (1998). The Blue Beach Member was interpreted as a near-shore basin periodically connected to the sea. During the onset and peak of marine transgression diverse species of Fusulinina and Miliolina foraminifers were recorded in association with marginal marine ostracod species belonging to the Palaeocopida and Bairdiacea. With the gradual withdrawal of the marine influence a retreat of marine fusulinids and increasing dominance of agglutinated Textulariidae (mainly *Trochammina*, *Ammobaculites* and *Ammodiscus*) occurred. Under low salinity conditions the ostracod assemblage also changed. Marine species were replaced with euryhaline *Paraparchites* and *Cavellina*, the brackish *Geisina*, and by freshwater species of *Carbonita*.

In the Devonian to Permian deposits of the Brabant Massif in Belgium the shallow near-shore environment was inhabited by a mixed marine ostracod assemblage represented mainly by rare bairdiacean ostracods (notably *Bairdia* and *Acratia*) (Bless et al. 1988). However, because of the absence of fusulinid foraminiferans this environment was interpreted as "less open marine environment". The brackish water setting of the deltaic system under marine incursion was inhabited by brackish *Geisina* and freshwater *Carbonita* species associated with agglutinated forms of Foraminifera (*Ammodiscus* and *Hyperammina*). Generally, the majority of Early Carboniferous fossil evidences of both, agglutinated and Fusulinida Foraminifera come from North America, Canada and Western Europe, with fewer records from Asia and Australia (Fig. 1).

In modern environments a similar pattern is observed. Hedberg (1934) recorded arenaceous (agglutinated) and calcareous Foraminifera in freshwater habitats in Venezuela. Foraminifera were collected from Lake Maracaibo which is connected with Caribbean Sea by a narrow neck, although salinity of this water body was low ($\approx$1 ‰). In freshwater habitats Hedberg (1934) found three groups of Foraminifera: agglutinated (arenaceous) forms belonging to the Lituolidae, Trochamminidae and Textulariidae; pelagic and benthic forms (*Globigerina*, *Bulimina* and *Uvigerina*), the occurrence of these marine species was viewed as a result of transportation by tidal currents; and euryhaline species (*Rotalia beccari*, *Elphidium* sp.) with *Quinqueloculina fusca* as a dominant species. *Quinqueloculina fusca* has also been reported in Tertiary sediments together with brackish water molluscs as an associated group, and in the absence of marine fauna.

Holzmann and Pawlowski (2002) found naked forms of allogromids in Lake Geneva and two other freshwater bodies in Switzerland. Authors obtained foraminiferal DNA sequences which clustered with a clade of saccamminid Foraminifera. Holzmann and Pawlowski also examined the 19th-century collection of Eugene Penard. Five freshwater species of Penard's *Gromia* could be distinguished according to type of agglutinated test (covered with small, siliceous particles).

Fig. 1 Distribution of Carboniferous Foraminifera referred to in this article. The palaeogeographic map is a reconstruction of the Pennsylvanian 300 Ma modified from Ron Blakey (http://cpgeosystems.com/mollglobe.html). Numbers next to Fusulinida (open circles) and agglutinated foraminifera (dots) occurrences indicate where the record is published in the literature: 1: North America, Sahul Shelf, Payne et al. 2012. 2: Wyoming, USA, Mamet 1975. 3: Nova Scotia, Canada, Wightman et al. 1994. 4: Nova Scotia, Canada, Tibert and Scott 1999. 5: Ireland, Gallagher 1998. 6: Austria, Krainer et al. 2002. 7: France, Conil et al. 1986. 8: Italy, Krainer and Davydov 1998. 9: Austria, Italy, Slovenia, Russia (Southern Ural) Forke 2002. 10: Czech Republic, Kalvoda 2002. 11: Iran, Leven and Gorgij 2011. 12: Iran, Afghanistan, Turkmenistan, Leven 2010. 13: South China, Ke-liang 1987. 14: Northern Australia, Mamet and Belford 1968

Typically, crossing from a brackish to freshwater regime is characterised by the disappearance of Foraminifera other than organic-walled Allogromiidae (Sen Gupta 2003) or/and other agglutinated forms. Most allogromids are widespread in marine environments, but some genera have been described from freshwater or even terrestrial habitats (Meisterfeld et al. 2001; Lejzerowicz et al. 2010). The allogromid test is usually membranous or proteinaceous, which in some cases, may be covered with agglutinated foreign particles (Loeblich and Tapan 1987; Holzmann and Pawlowski 2002). In ostracods the main difference between non-marine and marine carapace is the proportion of calcium carbonate and organic components. In freshwater species the content of $CaCO_3$ is significantly lower and valves consist of mainly chitinous fibres (Keyser and Walter 2004).

Agglutinated Foraminifera are defined according to the specific structure of their test. The cement that binds the wall of the test together may be organic (e.g. Astrorhizida), calcareous (e.g. Textulariida) or of a mixed nature (e.g. Lituolida) containing organically-cemented, calcareous and microgranular types (Kaminski 2004). Agglutinated species build their test using ambient components. Many species do not select or only weakly select the grains that are incorporated into their test (Thomsen and Rasmusen 2008), but some may be strongly selective (Murray 1963). Grains used by Foraminifera are composed mainly of feldspar (70 % plagioclase and 30 % alkali orthoclase). Other grains comprise pure silica and Ca, Fe, Mg-silicates (du Châtelet et al. 2013). In comparison, Cambrian agglutinated Foraminifera consist mainly of microgranular quartz, with inclusions of feldspar, clay minerals and opaque minerals (McIlroy et al. 2001). Du Châtelet et al. (2013) observed that proportion of quartz used by Foraminifera was species dependent and detected only a single grain of calcite. In general, agglutinated Foraminifera are not commonly regarded as calcifying, although, some species (e.g. *Textularia oviedoiana*) produce a low Mg calcitic matrix comparable in composition to the needles of miliolids (Bender and Hemleben 1988).

3 Physiological Adaptations

3.1 Feeding Strategy

Ostracods exhibit a wide variety of feeding strategies. Large planktonic ostracods usually are active predators, although it is still unclear whether they catch living or/and dead prey/food items (Vannier et al. 1998). Other pelagic halocyprids probably feed on suspended material: these are likely to be microphagous feeders utilizing foraminifers, diatoms, silicoflagellates and detrital aggregates as well (Angel 1990). Other myodocopids, which are mainly benthic dwellers, are either filter-feeders (e.g. cylindroleberidinids: Cannon 1933), carnivorous scavengers (e.g. cypridinids: Cohen 1983; Vannier and Abe 1993; Vannier et al. 1998), detritus feeders (e.g. philomedids: Hartmann 1975) and predators on small invertebrates (e.g. rutidermatids: Cohen and Kornicker 1987).

Ostracods display numerous morphological adaptations for their feeding strategy. Adaptation to scavenging and predation is reflected in the morphology of their mandibles, maxillulae (4th limbs), maxillae (5th limbs), and furcal lamellae. Powerful furca are used as a gripping tool to hold moving prey or to anchor onto a dead prey item. The feeding appendages are involved in capturing and physical breakdown of food, passing food pieces into the mouth, tearing off lumps of prey soft body and biting (Vannier et al. 1998). Planktonic halocyprids use their furca for selective rejection of food items from within the carapace (Lochhead 1968; Vannier et al. 1998). Filter-feeding species developed large vibratory plates that generate continuous flow fields across the body that serve a respiratory as well as feeding function. Mandibles of freshwater podocopids consist of masticatory processes with strong teeth and well developed vibratory plates. Masticatory processes associated with the lower lip form so-called food-rakes which assist the mandibles in breaking down food and passing it into the mouth (Meisch 2000). The parasitic form *Sheina orri* (Cypridinidae) uses its mandibular and maxillar claws to anchor itself to the gill tissues of its shark host (Bennett et al. 1997; Vannier et al. 1998).

As a group, the Foraminifera utilise also a broad range of feeding strategies—from osmotrophy to various holotrophic mechanisms (Pawlowski et al. 2003). Carboniferous fusulinids were most probably active herbivores, detritivores and omnivores (Gallagher 1998). Modern foraminifers acquire nutrients through direct uptake of DOC, suspension feeding, grazing, deposit feeding, predation, symbiosis and parasitism (Goldstein 2003). Suspension feeding is common mainly in pelagic forms, but it is worth noting that they do not have mechanism for creating water currents as ostracods do, thus, Foraminifera are probably less efficient as 'passive' suspension feeders. The pseudopodia of carnivorous species are adapted for capturing prey. Many Foraminifera are not strictly carnivorous, but utilise some other feeding strategies. Several species lead a parasitic mode of life, for example on bivalve molluscs, sponges, stone corals, or even other foraminiferans as a host. Symbiotic relationships in the Foraminifera include algal endosymbiosis, chloroplast husbandry (kleptoplastidy) and bacterial endosymbiosis. In general, the symbiont supplies the foraminifer with organic nutrition, and the host, in return, provides the symbiont with a fairly stable microenvironment and with other compounds (dissolved nitrogen, phosphorus, etc.) (Goldstein 2003). Possessing endosymbionts is a beneficial adaptation in colonisation of new habitat. Foraminifera in a symbiotic relationship have independent source of organic carbon, although the majority of symbionts are light dependent.

Eukaryotic protozoans acquire food by way of endocytic uptake and subsequent intracellular digestion within a discontinuous system of vacuoles. Bowser et al. (1985) in their experiment on *Allogromia* species observed that the cytoplasm contains only discrete vacuoles, and there is no evidence for a presence of 'primitive gut' or lacunary system. During their experiments some of observed vesicles broke, which Bowser et al. (1985) inferred was due to hypo-osmotic shock. It appears that similar looking vacuoles may be involved in diverse physiological processes.

The best feeding strategy for the colonisation of a new habitat would appear to be generalist omnivory and detritivory, so that organisms will find food in every aquatic habitat. Indeed, the survival of first ostracod freshwater invaders may have been dictated by their feeding strategy. In the Pennsylvanian Coal Measures of northern England *Carbonita*, thought to be a deposit feeder, was interpreted as better adapted and more successful in its freshwater invasions than *Geisina* (filter feeder) (Bless and Pollard 1973; Bennett 2008). Both groups, Foraminifera and Ostracoda, include omnivorous and detritivorous species. Feeding strategy was unlikely to be a barrier for the efficient colonisation of fresh water in either of these taxa.

3.2 Osmoregulation

The majority of marine ostracods appear to be isotonic with ambient sea water, but all freshwater forms must be osmoregulators. In inland waters organisms are hyperosmotic regulators, while dwellers of hyperhaline environments must be hypoosmotic regulators (Lockwood 1962).

The freshwater medium is hypotonic to body fluids, so ostracods need to take up salt from their food and/or absorb salts in their antennal glands. Embryos utilise special cells located in the non-calcified zone of the inner valve layer for salt reabsorption (Aladin and Potts 1996). Hypoosmotic regulation in both adults and juveniles may be achieved by drinking the medium and by the excretion of salts by cells in the inner, non-calcified shell layer. The cuticle of this zone is characterised by high permeability to ions, and salt excretion seems to be under strict control of 'caplike structures', which are most likely salt glands (Aladin 1983, 1984, 1993; Aladin and Potts 1996).

Tolerant, euryhaline species are capable of adapting to changing salinity conditions. In the Australian euryhaline species *Mytilocypris praenuncia* Aladin and Potts (1996) observed changes in external morphology depending on salinity: below 4 ‰ cells had clear borders and there were numerous depressions in the cuticle but between 8 and 12 ‰ the cell borders and holes in the cuticle vanished. Finally, when *M. praenuncia* was raised in salinities from 20–34 to 44–48 ‰ the cells regained clear borders, and salt glands appeared. It is worth noting that all these changes in morphology occurred only during moulting, so the physiological adaptations from one level of osmoregulation to the next can be completed only during a moult. Aladin and Potts (1996) also noticed that after moulting haemolymph concentration quickly returns to its previous state, and the time taken to reach equilibrium varied from 2 to 26 h.

There are few studies on the effect of salinity on foraminiferal physiology under experimental conditions. Murray (1963) observed that the benthic species *Elphidium crispum* thrived in water at 30–35 ‰, survived in 25 ‰ and died in salinities below 25 ‰. Specimens which had been exposed to low salinity for several days quickly resumed normal feeding rates when returned to normal sea water and had apparently not been permanently affected by the unfavourable conditions. Murray (1963) also noticed that *E. crispum* was capable of tolerating lower salinity for a few weeks (38 days), and survival was better in lower temperature. Most probably the main effect of lowering temperature is the slowing of the rate of metabolic processes. In elevated salinity (50 ‰) Murray observed retarded growth and inhibited reproduction, but exposure did not result in death. This observation was confirmed by Bradshaw (1955) in *Rotaliella heterocaryotica*, who observed active growth when salinity reached 23.5 ‰ and a cessation in too high (37 ‰) and in too low (16.8–20.1 ‰) salinities.

Osmoregulation in ostracods seems to be more specialised than in Foraminifera. Ostracods possess specialised cells, tissues and glands which can provide effective osmoregulation over wide range of salinity (0–70 ‰). Additionally, crustaceans in general are known for their ability to reduce membrane permeability (Lee and Bell 1999). Foraminifera also seem to be adaptable in regard to changeable salinity conditions. Typically they are equipped with numerous vacuoles, which are used in various physiological processes. It seems to be probable that some of them are not strictly specialised, and can change their role with changing environmental conditions. Undoubtedly, another important factor in osmoregulation is the permeability of the test and body membrane, but this should be confirmed by further investigations.

3.3 Reproduction

The overwhelming majority of marine ostracods exhibit sexual reproduction, although some brackish and freshwater species have acquired the ability to reproduce asexually. "(…) sexual propagation may be regarded as a source of individual variability, furnishing material for the operation of natural selection" (August Weismann 1887, quoted in Butlin et al. 1998a). Apart from undoubted genetic benefits of sexual reproduction there are some costs. The main ones are the 'cost of males', the requirement for males in sexual lineages, and the 'cost of mating', the energetic costs of finding partner, courting, copulation, and predation and disease risk involved in mating (Butlin et al. 1998a).

It is still uncertain why parthenogenesis is so common in non-marine ostracods. There are a few different forms of ostracod asexuality (Butlin et al. 1998b; Martens 1998). An ancient asexual does not have close sexual relatives, and its populations consist exclusively of females. Some lineages have geographically restricted sexual and asexual populations—so-called 'geographical parthenogenesis'. Finally, there are populations whose sexual and asexual lineages coexist. The last mode, referred to as 'mixed reproduction', seems to be the most beneficial for new habitat colonisation. Parthenogenesis enables quick and easy dispersion and increase in abundance, because a single egg is sufficient to invade a new water habitat. Alternatively, in highly changeable brackish and freshwater habitats, the diversity of sexual parents offspring might have a better chance to adapt and survive. The first ostracods recorded as freshwater invaders in the Carboniferous probably exhibited mixed reproduction (Griffiths and Horne 1998; Liebau 2005; Bennett 2008).

Both fossil and modern ostracods can exhibit advanced reproductive strategies that facilitate survival in new salinity regimes. The production of resting eggs, or resistance to desiccation or other unfavourable environmental conditions, may also have been attributes of the first fresh water invaders, for example species of *Carbonita* from a temporary pond habitat, found in the Montceau Lagerstätte (Vannier et al. 2003). Brooding may also have allowed the colonisation of more extreme habitats. In the deposits from Lower Silurian Herefordshire Konservat-Lagerstätte myodocopan species with eggs and possibly juveniles were preserved, thus providing an unequivocal view of parental brood care as a reproductive strategy which has lasted within this group from the Silurian to present day (Siveter 2008).

The typical foraminiferal life cycle is characterised by an alteration of asexual and sexual generations. In the sexual generation the adult gamont produces gametes, and fertilisation takes place by the fusion of two gametes, usually from different parents. The zygote may spend a brief phase as a shell-less (naked) amoeba. In metazoans meiosis typically occurs during gametogenesis, however, in the asexual generation the foraminiferal agamont produces numerous offspring by multiple fission, with meiosis as an integral part of this process. Thus, haploid young individuals typically grow to become adult gamonts, which produce gametes by mitotic nuclear divisions (Pawlowski 2009).

Extant Foraminifera are known to exhibit numerous variations on this general cycle, however life cycles of only about 30 of over 10,000 modern species have been studied (Goldstein 2003). The variation includes trimorphism, apogamic life cycle, binary fission, various forms of budding, the occurrence of test and nuclear dimorphism. The alternation of generations in Foraminifera may by facultative or obligatory (Goldstein 2003).

Some species are apogamic and they have reduced the complexity of their life cycle by omitting the sexual generation. In *Fissurina marginata* and *Spiroloculina hyalina* only the asexual phase is observed (Arnold 1964), whereas some planktonic species reproduce exclusively sexually, and no asexual generation has been observed (Goldestein 2003).

The life cycle of allogromiids and astrorhizids seems to be more variable than in other Foraminifera. Binary fission has been observed in *Allogromia laticollaris*, budding, serial and multiple budding in *Saccammina sphaerica*, *S. alba*, and *A. laticollaris*, fragmentation occurs in the miliolid *Calcituba polymorpha*, and *Floresina* is capable of producing multiple broods (Arnold 1954, 1964, 1967; Goldstein 1988, 2003).

Most groups of unicellular organisms (including most Foraminifera) adopt an opportunistic r-strategy, where the population grows quickly by frequent cell divisions. However, there is some evidence that larger-sized benthic species of Foraminifera conform to a K-selected mode of life over long periods. The production of a complex, large-sized test which is slow growing, houses storage products and symbionts, and generates the permanent body shape can be viewed as advanced adaptations to their particular mode of life (Hottinger 1982).

For benthic Foraminifera there are four methods of dispersal: (1) Release gametes, zygotes, or embryonic agamonts or gamonts into the water column, (2) Meroplanktonic juvenile stages with subsequent passive spread by currents, (3) Self-locomotion along sea floor, and (4) Passive dispersal by means of a physical or biological vector (Alve 1999). Sexual generation appears to be the most efficient method, because released gametes are advected by the bottom water currents (Kitazato and Matsushita 1996). Dispersal by gametes may be efficient over short distance, whereas zygotes and embryonic juveniles, with their density comparable to sea water, are more prone to disperse over larger distance (Alve 1999).

3.4 Other Adaptations

Foraminifera show additional adaptations against unfavourable environmental conditions. They are traditionally considered to be obligate aerobes, and most species become dormant during exposure to adverse conditions such as oxygen depletion. However, some benthic allogromiid species living in low-oxic habitats are capable of storing and respiring nitrate through complete denitrification to N_2 (Kuhnt et al. 2013). Foraminifera have evolved at least two ways to carry out this process: one involving symbionts and the other by the foraminifer itself. Bernhard et al. (2012) observed that species capable of denitrification possess large vacuoles containing sea water with a high concentration of nitrate. Recent studies revealed that the gene for nitrate reduction could be localised to the symbiont or to the allogromiid. Foraminifera produced N_2 from NO_3^- and rapidly consumed intra-cellular nitrate during both oxic and anoxic incubations, thus denitrifying species should be regarded as facultative anaerobes (Kuhnt et al. 2013), although it was observed that presence of oxygen partially inhibited or delayed the onset of nitrate respiration (Bernhard et al. 2012). In oxygen-depleted environments anaerobic metabolic pathways are required. In the Proterozoic or Palaeozoic, during oxygen crises, the ability to denitrify could have imparted a major ecological advantage and contributed to the success of early foraminiferan lineages.

The rate of colonisation depends in part upon the geochemical characteristics of the new habitat. Substrates which were previously anoxic have completely different physical and chemical properties compared to well-oxygenated environment. Alve (1995) recorded that for opportunistic species of Foraminifera it took more than 1 year to colonise sediment which had experienced 5 years of anoxia, however, invasion by less efficient species may take several years (Alve 1999). It seems that in marine environment Foraminifera are able to invade more efficiently than macrofaunal invertebrates. Kaminski et al. (1988) observed for agglutinated Foraminifera that 9 months may be sufficient time to recover to background levels of diversity and abundance after severe disturbance (during which time, the macrofauna did not recover).

All studied physiological adaptations developed by Foraminifera and Ostracoda are collated and summarized in Table 1.

4 Calcification

Both groups, Ostracoda and Foraminifera, seem to be successful invaders of new environments. They adopted a broad range of feeding strategies and reproduction modes. The production of resting stages and brood care may also have contributed to them being efficient invaders. They are also both highly tolerant to variations in salinity. Foraminifera and Ostracoda have the ability to construct saturated with calcium carbonate exoskeleton. How do organisms with a high demand for calcium compounds cope with the low availability of this element in freshwater habitat?

The shell of ostracods is important for protection, respiration, metabolism and osmoregulation (Okada 1982; Keyser 1990; Aladin 1993). Sohn (1958) reported the following constituents for the shell of *Chlamydotheca unispinosa*: 82.7 % calcium carbonate, 12.8 % protein, 2.2 % chitin and 1.9 % trace elements as K, Mg, Na, Si, Sr, Al and Ba. Sohn (1958) also noticed that the $CaCO_3$ content is variable depending on species, and ranges from 80 to 90 %. The cuticle of the shell is mineralised with low magnesium calcium carbonate in the form of calcite, but never contains aragonite. The highest content of calcium carbonate is found in the shell of marine Cytheroidea, whereas in freshwater forms, as *Cypria ophtalmica*, the

Table 1 Physiological adaptations of Foraminifera and Ostracoda in marine and freshwater environments (see text)

	Marine	Freshwater		
Adaptation	Foraminifera	Ostracoda	Foraminifera	Ostracoda
Feeding strategy	Predators, scavengers, filter-feeders, herbivorous, omnivorous, detritivorous, DOC feeding, symbiotic, parasitic[1])	Predators, scavengers, filter-feeders, omnivorous, detritivorous, symbiotic, parasitic[2])	Herbivorous[3]), others?	Omnivorous, detritivorous, filter-feeders[4])
Osmoregulation	≈25-35 ‰, retarded growth and inhibited reproduction in lower and higher salinity; isotonic with ambient sea water[5])	Tolerant, depending on species range up to 0-70 ‰; isotonic with ambient sea water[6])	0-? ‰; contractile vacuoles for osmoregulation3)	0-8‰; 'cap-like structures' (salt glands) for osmoregulation[6])
Reproductive strategy	Alteration of sexual and asexual generations, budding, fragmentation[1])	Almost exclusively sexual reproduction, brood care[7])	Asexual reproduction, binary fission[8]), others?	Sexual, asexual and mixed reproduction, brood care[4], [7])
Calcification	The higher salinity the more calcified test[9])	The higher salinity the more calcified valves[10])	Almost exclusively agglutinated, organic-shelled or naked species[3], [8], [11])	Thin, less calcified and poorly ornamented carapace[4], [12])
Other adaptations	Resting stages, diapause, nitrate respiring in anoxic environment[13])	Resting stages[14])	Resting stages, diapause[8])	Resting stages, resistance to desiccation, diapause[4])

1: Goldstein 2003
2: Vannier et al. 1998
3: Holzmann and Pawlowski 2002
4: Meisch 2000
5: Murray 1963
6: Aladin and Potts 1996
7: Martens 1998
8: Meisterfeld et al. 2001
9: Dueñas-Bohórquez et al. 2009
10: Chivas et al. 1986
11: Sen Gupta 2003
12: Keyser and Walter 2004
13: Kuhnt et al. 2013
14: Vannier and Abe 1992

$CaCO_3$ content is significantly lower. In habitats with low calcium content they cannot construct fully calcified carapace and the valve may then consist primarily of chitinous fibres (Keyser and Walter 2004). In low salinity environments ostracods seem to have less mineralised, mainly chitinous valves, but they still have carapace to protect their soft body parts.

The ostracod carapace is shed by moulting up to eight times during development and each stage has new and more heavily calcified valves. In ostracods the calcite is not reabsorbed from the old carapace during moulting, as happens in many malacostracan crustaceans, but is removed and formed again during calcification of the new valves (Turpen and Angell 1971; Keyser and Walter 2004). Prior to moulting ostracods begin producing the carapace by the absorption of a large amount of calcium compounds and chitin precursors. The uncalcified inner lamella cuticle is formed by the inner epidermal cells (Yamada and Keyser 2010). The outer epidermal layer beneath the calcified cuticle contains large amounts of granules within the cells. These intracellular bodies contain compounds of calcium phosphate and small amounts of sodium, potassium, chloride and sulphur. Neither magnesium nor strontium (known to be present in the fully calcified carapace) is found in these granules. In the next step calcium is released from the globules, penetrates the epidermal membrane and then forms granules of amorphous calcite outside the membrane. Some species (e.g. *C. ophtalmica*) retain amorphous calcite in the carapace, but most transform the calcite into the final crystalline arrangement in the epidermal layer, because amorphous calcite dissolves easily. In the juvenile stages crystallisation is not complete and the organisms have weaker shells. This mechanism is similar for both marine and freshwater ostracods (Turpen and Angell 1971; Keyser and Walter 2004; Yamada et al. 2005; Yamada and Keyser 2010).

The chemistry of the ostracod valve is a function of the surrounding water chemistry modified by temperature, calcification rate, and inter- and intra-specific variability (Van der Meeren et al. 2011). The influence of salinity levels on ostracod calcification is unclear. Some authors (e.g. De Deckker et al. 1999) found no relationship between Mg/Ca content of the ostracod valve and salinity, while others (e.g. Chivas et al. 1986) found a positive correlation between salinity and Mg/Ca content in the carapace. Decrouy et al. (2011) noticed that in shallow waters higher temperatures increased the Mg/Ca and DIC concentration of water, which may have an effect on ostracod mineralisation. Additionally, Carbonel et al. (1988) suggested that the structure, ornamentation and size of the carapace may be correlated with the degree of salinity.

Eleven out of 15 extant orders of Foraminifera precipitate calcareous tests and thus are among the major producers of calcium carbonate in the oceans (Hansen 2003; Bentov et al. 2009). In pelagic foraminifers the test wall consists of extremely pure calcite (about 99 % by weight $CaCO_3$) and trace elements such as Mg, Sr, Ba and Cd. Elements are incorporated directly from ambient sea water during test precipitation, thus shell composition reflects chemical composition of the medium, and both physical and biological conditions present during calcification (Lea 2003).

According to test structure calcifying Foraminifera are commonly divided into two groups: miliolid and hyaline. The miliolid test contains relatively high Mg/Ca ratios and the hyaline test has much lower ratios of Mg/Ca (Vogel and Uthicke 2012). Miliolids precipitate calcite in the form of 2–3 μm needles within cytoplasmic vesicles (Berthold 1976). These needles are accumulated within the cell and then they form a new chamber after simultaneous transport outside the test and assembly within the organic matrix (Angell 1980). The outer layer of the wall is arranged in dense rows of needles, to what gives a porcelaneous structure of the test surface (de Nooijer et al. 2009).

Foraminiferal calcification is preceded by extraction of calcium and bicarbonate ions from sea water. Hyaline species tend to store calcium and carbonate in separate intracellular organelles. During calcification Foraminifera build their new calcite chamber over their previous shell (Bentov et al. 2009). Chamber formation starts with the production of a primary organic sheet (POS) (de Nooijer et al. 2009). During formation of the POS cytoplasm with a raised pH (≥9.0) is transported to the site of calcification. Vesicles with high pH are formed mainly in the penultimate chamber and transported through the ultimate chamber to its aperture where calcification occurs. Most probably protons are pumped out from the vesicles and stored in a specialised cytosolic compartment with low pH (≤6.0). Throughout chamber formation vesicles with elevated pH are continuously transported to the calcification site until chamber formation is complete (de Nooijer et al. 2009).

Miliolids also need pH ≈ 9.0 for calcification, but their high pH vesicles are conveyed around in the cytoplasm relatively fast and in a seemingly undirected manner. However, the vesicles containing calcitic needles have considerably lower pH (7.5–8.0). Elevating the pH is a widespread strategy to promote calcite precipitation in Foraminifera: it overcomes the inhibition by Mg^{2+} of calcite precipitation that prevents spontaneous crystal nucleation and growth in modern seawater Mg/Ca ratios (Zeebe and Sanyal 2002), and it also promotes the conversion of bicarbonate into carbonates (Zeebe and Wolf-Gladrow 2001).

Most models of biomineralisation assume involvement of membrane ion transporters (channels and pumps) for delivery of Ca^{2+} and other ions to the calcification site. However, Bentov et al. (2009) observed another mechanism in the shallow water, benthic foraminiferan *Amphistegina lobifera* (hyaline), in which transport of vacuoles with sea water via fluid phase endocytosis may account for most of the calcium and other ions. Initially, vacuoles are semi open to external sea water and filling of the vacuoles may be mediated by narrow, tubular channels. During intracellular endocytosis sea water vacuoles undergo alkalisation and this further enhances their calcifying potential. The alkalinisation of the vacuoles suggests that the supply of CO_3^{2-} for calcification is also mediated by the sea water vacuoles (Ferguson et al. 2008). The massive calcium transport through the cytosol would require a large expenditure of energy. In addition, because of the lower cytoplasm pH (7.2–7.5), direct transport through the cytoplasm may hinder the maintenance of the high pH needed for calcification. Vacuolar transport obviously reduces these problems, and can bring Ca^{2+}-enriched solution to the calcification site bypassing the cytoplasm (Bentov et al. 2009). De Nooijer et al. (2008, 2009) recorded that the phenomenon of alkaline vacuole (vesicles) is general, encompassing both hyaline and miliolid species. Thus, sea water seems to be a calcifying solution, in agreement with the biomineralisation model of Elderfield et al. (1996). Elderfield et al. (1996) observed that precipitation occurs at the mineralisation site, which is isolated from the outside medium, however, the medium supplies the extraneous ions for precipitation. The biomineralisation reservoir is similar but not necessarily identical to sea water in composition. Sea water provides chemicals for calcification and controls the diffusion gradient, and thus influences the test composition. Organisms probably extract calcium from vacuoles with sea water into the storage organelles (Elderfield et al. 1996).

According to Hottinger 2000) new chamber formation in Foraminifera includes two processes: rhizopodial extrusion and biomineralisation. It seems that the rhizopodia play a key role in a test formation. A new chamber is built on a rhizopodial skeleton formed by microtubules. In the next stage rhizopodial skeleton is saturated with skeleton elements in a biomineralisation process (Hottinger 2000). Most probably in a low mineralisation environment they cannot complete the precipitation of new test, but the naked rhizopodial skeleton is not hard enough to protect them against predators and other unfavourable environmental conditions.

The following parameters seem to be involved in the control of foraminiferal Mg/Ca and Sr/Ca ratios: water temperature, salinity, calcite saturation, carbonate ions concentration, water pressure, ontogeny and growth rate (Elderfield et al. 1996; Lea 2003; Dueñas-Bohórquez et al. 2009, 2011). An impact of salinity on foraminiferal calcification has been observed in several studies (e.g. van Raden et al. 2011). According to Dueñas-Bohórquez et al. (2009) the average Mg/Ca values of planktonic Foraminifera *Globigerinoides sacculifer* increase in higher salinities despite the relatively large inter-individual variability. Ferguson et al. (2008) suggested that salinity is the most likely environmental factor to explain unusually high Mg/Ca ratios of Mediterranean Foraminifera. They recorded that correlations of Mg/Ca with the salinity at which organisms calcified were more highly significant than those with calcification temperatures.

5 Discussion

In general, both Foraminifera and ostracods are highly adaptable and efficient colonisers of new aquatic habitats. A key to success for the Foraminifera has been that they are not highly specialised. Leven (2010) claims that the reason why some species of Fusulinida (e.g. *Pseudostaffella*) became extinct in the Middle Carboniferous, was because they attained too high a degree of specialisation. In general, Foraminifera as simple, unicellular organisms maintained high plasticity and the ability to adapt to changing environmental conditions. Ostracods are more specialised but they have flexible genetic systems which allow species to readily adapt to the local environment (Carbonel et al. 1988).

Calcification in Foraminifera probably appeared during the Early Cambrian radiation, when miliolid and agglutinated Foraminifera separated from each other (Pawlowski et al. 2003). Building the test was one of the key adaptations for the initial diversification of Foraminifera. The test provided some protection against adverse environmental conditions and predation, as well as compartments in which they can store food, protect juveniles and house symbionts. For ostracods the development of a fully calcified carapace also played key role in their rapid diversification. At the turn of the Cambrian and Ordovician the majority of weakly-calcified ostracodomorphs (bradoriids sensu lato, Phosphatocopida) were replaced by well-calcified podocopomorphs (Liebau 2005).

Along the salinity gradient from a river to the sea, body size increases with increasing salinity (Gunter 1947). Indeed, animals inhabiting estuaries are smaller than their marine counterparts. The factor which is most likely to be responsible for small size of both Foraminifera and ostracods is the availability of calcium. Sea water salinity 35 ‰ contains about 400 ppm of calcium, whereas in comparison the hardest of river waters contains almost negligible amounts (Murray 1963). Recent experiments on Foraminifera revealed poor efficiency of calcium utilisation—only about 30 % of available Ca is used for shell formation (Böhm et al. 2012). It seems most likely that the barrier against massive colonisation of the freshwater realm by Foraminifera is their inability to construct a fully calcified test in low salinity regimes. The mechanism of foraminiferal calcification is strictly dependent on salt water content. They use sea water not only as a source of ions to construct their shell, but also as a biomineralisation solution, thus Foraminifera are typically defined as marine organisms (Holzmann et al. 2003). Only few genera are occasionally represented by species in low salinity environments, and the overwhelming majority of them are agglutinated or organic-shelled forms, which do not produce, or produce only small amounts of calcite. The success of ostracods in fresh water can be attributed to their development of a more effective mechanism of calcification. Less calcified ostracods are still sheltered by chitinous valves, but calcifying Foraminifera without biomineralisation process most likely are completely defenceless with their rhizopodial skeleton exposed. In low salinity environments ostracods construct less calcified, thinner and poorly ornamented valves, but they are still able to build a complete hard shell.

However, there is still a problematic question about the role of cellular organisation level. Ostracoda, as a multicellular group, seem to be better adaptable, than Foraminifera, their distant unicellular relatives. It is possible that tissue- and organ-level organisation has a greater capacity for physiological adaptation to new salinity regime than unicellular organisation does. That issue requires more attention in further studies.

Acknowledgments We are very grateful to Prof. Jan Marcin Węsławski, Prof. Marek Zajączkowski (Institute of Oceanology Polish Academy of Sciences, Sopot) and Prof. Geoffrey Boxshall (Natural History Museum, London) for their constructive and helpful comments on the manuscript.

References

Aladin NV (1983) Salinity adaptations and osmoregulatory abilities of the Ostracoda from the Caspian and Aral seas and the Brachiopoda and Ostracoda from the Caspian and Aral seas. Zoologichesky Zh 62:51–57

Aladin NV (1984) Salinity adaptations and osmoregulation abilities of Ostracoda from Black and Azov seas. Zoologichesky Zh 63:185–190

Aladin NV (1993) Salinity tolerance, morphology and physiology of the osmoregulatory organ in Ostracoda with special reference to Ostracoda from the Aral Sea. In: McKenzie KG, Jones PJ (eds) Ostracoda in the earth and life sciences. A.A. Balkema, Rotterdam, pp 387–403

Aladin NV, Potts WTW (1996) The osmoregulatory capacity of the Ostracoda. J Comp Physiol B 166:215–222

Alve E (1995) Benthic foraminiferal distribution and recolonization of formerly anoxic environments in Drammensfjord, southern Norway. Mar Micropaleontol 25:169–186

Alve E (1999) Colonization of new habitats by benthic foraminifera: a review. Earth Sci Rev 46:167–185

Angel MV (1990) Food in the deep ocean. In: Whatley R, Maybury C (eds) Ostracoda and global events. Chapman and Hall, London, pp 273–285

Angell RW (1980) Test morphogenesis (chamber formation) in the foraminifer *Spiroloculina hyalina* schulze. J Foramin Res 10:89–101

Arnold ZM (1954) Variation and isomorphism in *Allogromia laticollaris*: a clue to foraminiferal evolution. Contrib Cushman Found Foram Res 5:78–87

Arnold ZM (1964) Biological observations on the foraminifer *Spiroloculina hyaline* schulze. Univ Calif Publ Zool 72:1–93

Arnold ZM (1967) Biological observations on the foraminifer *Calcituba polymorpha* Roboz. Arch Protistenk 110:280–304

Bender H, Hemleben C (1988) Constructional aspects in test formation of some agglutinated foraminifera. Abhandlungen der geologischen Bundesanstalt 41:13–21

Bennett C (2008) A review of the Carboniferous colonisation of non-marine environments by ostracods. Senckenb Lethaea 88:37–46

Bennet CE, Siveter DJ, Davies SJ, Williams M, Wilkinson IP, Browne M, Miller CG (2012) Ostracods from freshwater and brackish environments of the Carboniferous of the Midland Valley of Scotland: the early colonization of terrestrial water bodies. Geol Mag 149:366–396

Bennett MB, Heupel MR, Bennett SM, Parker AR (1997) *Sheina orri* (Myodocopa: Cypridinidae) an ostracod parasitic on gills of the epaulette shark, *Hemiscyllium ocellatum* (Elasmobranchii: Hemiscyllidae). Int J Parasitol 27:275–281

Bentov S, Brownlee C, Erez J (2009) The role of seawater endocytosis in the biomineralization process in calcareous foraminifera. Proc Natl Acad Sci 106:21500–21504

Bernhard JM, Casciotti KL, McIlvin MR, Beaudoin DJ, Visscher PT, Edgcomb VP (2012) Potential importance of physiologically diverse benthic foraminifera in sedimentary nitrate storage and respiration. J Geophys Res 117:G03002

Berthold WU (1976) Biomineralisation bei milioliden Foraminiferen und die Matritzen-Hypothese. Naturwissenschaften 63:196–197

Bless MJM, Pollard JE (1973) Paleoecology and ostracode faunas of Westphalian Ostracode Bands from Limburg, The Netherlands and Lancashire, Great Britain. Meded Rijs Geol Dienst Nieuwe Serie 24:21–53

Bless MJM, Streel M, Becker G (1988) Distribution and paleoenvironment of Devonian to Permian ostracode assemblages in Belgium with reference to some Late Famennian to Permian marine nearshore to 'brackish-water' assemblages dated by miospores. Ann Soc Géol Belg 110:347–362

Böhm F, Eisenhauer A, Tang J, Dietzel M, Krabbenhöft A, Kisakürek B, Horn C (2012) Strontium isotope fractionation of planktic foraminifera and inorganic calcite. Geochim Cosmochim Acta 93:300–314

Boudagher-Fadel MK (2008) The Palaeozoic larger benthic foraminifera: the Carboniferous and Permian. In: Boudagher-Fadel MK (ed) Evolution and geological significance of larger benthic foraminifera. Elsevier Science, Amsterdam, pp 39–118
Bowser SS, McGee-Rusell SM, Rieder CR (1985) Digestion of prey in foraminifera is not anomalous: a correlation of light microscopic, cytochemical, and HVEM technics to study phagotrophy in two allogromiids. Tissue Cell 17:823–839
Bradshaw JS (1955) Preliminary laboratory experiments on ecology of foraminiferal populations. Micropaleontology 1:351–358
Butlin RK, Schön I, Griffiths HI (1998a) Introduction to reproductive modes. In: Martens K (ed) Sex and parthenogenesis: evolutionary ecology of reproductive modes in non-marine ostracods. Backhuys Publishers, Leiden, pp 1–24
Butlin RK, Schön I, Martens K (1998b) Asexual reproduction in nonmarine ostracods. Heredity 81:473–480
Calder JH (1998) The Carboniferous evolution of Nova Scottia. In: Blundell DJ, Scott AC (eds) Lyell: the past is the key to the present, vol 143. Geological Society, London, pp 261–302 Special Publications
Cannon HG (1933) On the feeding mechanism of certain marine Ostracoda. Trans Roy Soc Edinb 57:739–764
Carbonel P, Colin J-P, Danielopol D, Löffler H, Neustrueva I (1988) Palaeoecology of limnic ostracods: a review on some major topics. Palaeogeogr Palaeoclimatol Palaeoecol 62:413–416
Chivas AR, De Deckker P, Shelley JMG (1986) Magnesium content of non-marine ostracod shells: a new palaeosalinometer and palaeothermometer. Palaeogeogr Palaeoclimatol Palaeoecol 54:43–51
Cohen AC (1983) Rearing and postembryonic development of the myodocopid ostracode *Skogsbergia lerneri* from coral reefs of Belize and the Bahamas. J Crustac Biol 3:235–256
Cohen AC, Kornicker LS (1987) Catalog of the Rutidermatidae (Crustacea: Ostracoda). Smithson Contrib Zool 449:1–11
Conil R, Dreesen R, Lentz M-A, Lys M, Plodowski G (1986) The Devono-Carboniferous transition in the Franco-Belgian basin with reference to foraminifera and brachiopods. Anna Soc Géol Belg 109:19–26
Decrouy L, Vennemann TW, Ariztegui D (2011) Controls on ostracod valve geochemistry: part 1. Variations of environmental parameters in ostracod (micro-) habitats. Geochim Cosmochim Acta 75:7364–7379
De Deckker P, Chivas AR, Shelley MG (1999) Uptake of Mg and Sr in the euryhaline ostracods *Cyprideis* determined from *in vitro* experiments. Palaeogeogr Palaeoclimatol Palaeoecol 148:105–116
de Nooijer LJ, Toyofuku T, Oguri K, Nomaki H, Kitazato H (2008) Intracellular pH distribution in foraminifera determined by the fluorescent probe HPTS. Limnol Oceanogr Methods 6:610–618
de Nooijer LJ, Toyofuku T, Kitazato H (2009) Foraminifera promote calcification by elevating their intracellular pH. Proc Natl Acad Sci 106:15374–15378
du Châtelet EA, Bout-Roumazeilles V, Coccioni R, Frontalini F, Guillot F, Kaminski MA, Recourt P, Riboulleau A, Trentesaux A, Tribovillard N, Ventalon S (2013) Environmental control on shell structure and composition of agglutinated foraminifera along a proximal-distal transect in the Marmara Sea. Mar Geol 335:114–128
Dueñas-Bohórquez A, da Rocha RE, Kuroyanagi A, Bijma J, Reichart G-J (2009) Effect of salinity and seawater calcite saturation state on Mg and Sr incorporation in cultured planktonic foraminifera. Mar Micropaleontol 73:178–189
Dueñas-Bohórquez A, da Rocha RE, Kuroyanagi A, de Nooijer LJ, Bijma J, Reichart G-J (2011) Interindividual variability and ontogenetic effects on Mg and Sr incorporation in the planktonic foraminifer *Globigerinoides sacculifer*. Geochim Cosmochim Acta 75:520–532
Elderfield H, Bertram CJ, Erez J (1996) A biomineralization model for the incorporation of trace elements into foraminiferal calcium carbonate. Earth Planet Sci Lett 142:409–423
Ferguson JE, Henderson GM, Kucera M, Rickaby REM (2008) Systematic change of foraminiferal Mg/Ca ratios across a strong salinity gradient. Earth Planet Sci Lett 265:153–166

Forke HC (2002) Biostratigraphic subdivision and correlation of uppermost Carboniferous/lower permian sediments in the southern Alps: Fusulinoidean and Conodont faunas from the Carnic Alps (Austria/Italy), Karavanke Mountains (Slovenia), and southern Urals (Russia). Facies 4:201–276

Gallagher SJ (1998) Controls on the distribution of calcareous Foraminifera in the lower Carboniferous of Ireland. Mar Micropaleontol 34:187–211

Goldstein ST (1988) On the life cycle of *Saccamina alba* Hedley, 1962. J Foramin Res 18:311–325

Goldstein ST (2003) Foraminifera: a biological overview. In: Sen Gupta BK (ed) Modern Foraminifera. Kluwer Academic Publishers, New York, pp 37–56

Gray J (1988) Evolution of the freshwater ecosystem: the fossil record. Palaeogeogr Palaeoclimatol Palaeoecol 62:1–214

Grell KG (1973) Protozoology. Springen, Berlin

Griffiths HI, Horne DJ (1998) Fossil distribution of reproductive modes in non-marine ostracods. In: Martens K (ed) Sex and parthenogenesis: evolutionary ecology of reproductive modes in non-marine ostracods. Backhuys Publishers, Leiden, pp 101–118

Gunter G (1947) Extended remarks on relationships of marine animals to salinity. J Paleontol 21:498–500

Hansen HJ (2003) Shell construction in modern calcareous Foraminifera. In: Sen Gupta BK (ed) Modern Foraminifera. Kluwer Academic Publishers, New York, pp 57–70

Hartmann G (1975) Arthropoda, crustacea. 2. Buch, IV teil, 4. Lieferung. Bronn's Kl. Ordn Tierreichs, pp 572–786

Harvey THP, Vélez MI, Butterfield NJ (2012) Exceptionally preserved crustaceans from western Canada reveal a cryptic Cambrian radiation. Proc Natl Acad Sci 109:1589–1594

Hedberg HD (1934) Some recent and fossil brackish to fresh-water Foraminifera. J Paleontol 8:469–476

Holzmann M, Pawlowski J (2002) Freshwater foraminiferans from Lake Geneva: past and present. J Foramin Res 32:344–350

Holzmann M, Habura A, Giles H, Bowser SS, Pawlowski J (2003) Freshwater foraminiferans revealed by analysis of environmental DNA samples. J Eukaryot Microbiol 50:135–139

Hottinger L (1982) Larger Foraminifera, giant cells with a historical background. Naturwissenschaften 69:361–371

Hottinger L (2000) Functional morphology of benthic foraminiferal shells, envelopes of cells beyond measure. Micropaleontology 46:57–86

Kalvoda J (2002) Late Devonian-early Carboniferous foraminiferal fauna: zonation, evolutionary events, paleobiogeography, and tectonic implications. Masaryk Univ, Brno, pp 1–213

Kaminski MA (2004) The year 2000 classification of the Agglutinated Foraminifera. In: Bubik M, Kaminski MA (eds) Proceedings of the sixth international workshop on Agglutinated Foraminifera. Grzybowski Foundation Special Publication 8, pp 237–255

Kaminski MA, Grassle JF, Whitlatch RB (1988) Life history and recolonization among agglutinated foraminifera in the Panama basin. Abh Geol Bundesanst 41:229–243

Ke-liang W (1987) On the Devonian-Carboniferous boundary based on foraminiferal fauna from South China. Acta Micropaleontol Sin 2:002

Keyser D (1990) Morphological changes and function of the inner lamella layer of podocopid Ostracoda. In: Whatley R, Maybury C (eds) Ostracoda and global events. Chapman and Hall, London, pp 401–410

Keyser D, Walter R (2004) Calcification in ostracodes. Rev Esp Micropaleontol 36:1–11

Kitazato H, Matsushita S (1996) Laboratory observations of sexual and asexual reproduction of *Trochammina hadai* Uchio. Trans Proc Paleontol Soc Jpn New Ser 182:454–466

Krainer K, Vachard I, Vachard D, d'Ascq V (2002) Late Serpukhovian (Namurian A) microfacies and carbonate microfossils from the Carboniferous of Nötsch (Austria). Facies 46:1–26

Krainer K, Davydov V (1998) Facies and biostratigraphy of the Late Carboniferous/Early Permian sedimentary sequence in the Carnic Alps (Austria/Italy). Geoversitas 20:643–662

Kuhnt T, Friedrich O, Schmiedl G, Milker Y, Mackensen A, Lückge A (2013) Relationship between pore density in benthic foraminifera and bottom-water oxygen content. Deep Sea Res Part I Oceanogr Res Pap 76:85–95
Lea DW (2003) Trace elements in foraminiferal calcite. In: Sen Gupta BK (ed) Modern Foraminifera. Kluwer Academic Publishers, New York, pp 259–277
Lee CE, Bell MA (1999) Causes and consequences of recent freshwater invasions by saltwater animals. Tree 14:284–288
Lejzerowicz F, Pawlowski J, Fraissinet-Tachet L, Marmeisse R (2010) Molecular evidence for widespread occurrence of Foraminifera in soils. Environ Microbiol 12:2518–2526
Leven EJ (2010) Origin of higher fusulinids of the order Neoschwagerinida Minato et Honjo, 1966. Stratigr Geol Correl 18:290–297
Leven EJ, Gorgij MN (2011) First record of Gzhelian and Asselian Fusilinids from the Vazhnan formation (Sanandaj-Sirjan zone of Iran). Stratigr Geol Correl 19:486–501
Liebau A (2005) A revised classification of the higher taxa of the Ostracoda (Crustacea). Hydrobiologia 538:115–137
Lochhead JH (1968) The feeding and swimming of *Conchoecia* (Crustacea, Ostracoda). Biol Bull 134:456–464
Lockwood APM (1962) The osmoregulation of Crustacea. Biol Rev 37:257–305
Loeblich AJ, Tappan H (1987) Foraminiferal genera and their classification, vols 1–2. Van Nostrand Reinhold, New York
Mamet BL (1975) Carboniferous Foraminifera and algae of the Amsden formation (Mississippian and Pennsylvanian) of wyoming. US Government Printing Office, Washington
Mamet BL, Belford D (1968) Carboniferous Foraminifera, Bonaparte Gulf basin, Northwestern Australia. Micropaleontology 14:339–347
Martens K (1998) Sex and ostracods: a new synthesis. In: Martens K (ed) Sex and parthenogenesis: evolutionary ecology of reproductive modes in non-marine ostracods. Backhuys Publishers, Leiden, pp 295–321
Martens K, Schön I, Meisch C, Horne DJ (2008) Global diversity of ostracods (Ostracoda, Crustacea) in freshwater. Hydrobiologia 595:185–193
McIlroy D, Green OR, Brasier MD (2001) Palaeobiology and evolution of the earliest agglutinated Foraminifera: *Platysolenites*, *Spirosolenites* and related forms. Lethaia 34:13–29
Meisch C (2000) Freshwater Ostracoda of Western and Central Europe. In: Schwörbel J, Zwick P (eds) Süsswasser Fauna von Mitteleuropa 8(3). Spektrum Akademischer Verlag, Heidelberg
Meisterfeld R, Holzmann M, Pawlowski J (2001) Morphological and molecular characterization of a new terrestrial allogromiid species: *Edaphoallogromia australica* gen. et spec. nov. (Foraminifera) from Northern Queensland (Australia). Protist 152:185–192
Murray JW (1963) Ecological experiments on Foraminiferida. J Mar Biol Assoc U.K. 43:621–642
Newell ND (1949) Phyletic size increase, an important trend illustrated by fossil invertebrates. Evolution 3:103–124
Okada Y (1982) Ultrastructure and pattern of the carapace of *Bicornucythere bisanensis* (Ostracoda, Crustacea). Univ Mus Univ Tokyo Bull 20:229–255
Pawlowski J (2009) Foraminifera. In: Schaechter M (ed) Encyclopedia of microbiology. Elsevier, Oxford, pp 646–662
Pawlowski J, Holzmann M (2002) Molecular phylogeny of Foraminifera—a review. Eur J Protistology 38:1–10
Pawlowski J, Holzmann M, Berney C, Fahrni J, Gooday AJ, Cedhagen T, Habura A, Bowser SS (2003) The evolution of early Foraminifera. Proc Natl Acad Sci 100:11494–11498
Pawlowski J, Holzmann M, Tyszka J (2013) New supraordinal classification of Foraminifera: molecules meet morphology. Mar Micropaleontol 100:1–10
Payne JL, Boyer AG, Brown JH, Finnegan S, Kowalewski M, Krause RA, Lyons SK, McClain CR, McShea DW, Novack-Gottshall PM, Smith FA, Stempien JA, Wang SC (2009) Two-phase increase in the maximum size of life over 3.5 billion years reflects biological innovation and environmental opportunity. Proc Natl Acad Sci 106:24–27

Payne JL, Groves JR, Jost AB, Nguyen T, Moffitt SE, Hill TM, Skotheim JM (2012) Late Paleozoic Fusulinoidean gigantism driven by atmospheric hyperoxia. Evolution 66–9:2929–2939
Perrier V, Vannier J, Siveter DJ (2011) Silurian bolbozoids and cypridinids (Myodocopa) from Europe: pioneer pelagic ostracods. Palaeontology 54:1361–1391
Regier JC, Schultz JW, Kambic RE (2005) Pancrustacean phylogeny: hexapods are terrestrial crustaceans and maxillopods are not monophyletic. Proc Roy Soc B Biol Sci 272:395–401
Sen Gupta BK (2003) Foraminifera in marginal marine environments. In: Sen Gupta BK (ed) Modern Foraminifera. Kluwer Academic Publishers, New York, pp 141–160
Siveter DJ (2008) Ostracods in the Palaeozoic? Senckenb Lethaea 88:1–9
Siveter DJ, Briggs DE, Siveter DJ, Sutton MD (2010) An exceptionally preserved myodocopid ostracod from the Silurian of Herefordshire, UK. Proc Roy Soc B Biol Sci 277:1539–1544
Siveter DJ, Curry GB (1984) Lower Ordovician (Arenig) ostracods from the highland border complex. In: Curry GB, Bluck BJ, Burton CJ, Ingham JK, Siveter DJ, Williams A (eds) Age, evolution and tectonic history of the highland border complex, Scotland. Trans Roy Soc Edinb Earth Sci 75:113–133
Siveter DJ, Vannier JMC, Palmer D (1991) Silurian myodocopes: pioneer pelagic ostracods and the chronology of an ecological shift. J Micropalaeontology 10:151–173
Smith RJ (2000) Morphology and ontogeny of Cretaceous ostracods with preserved appendages from Brazil. Palaeontology 43:63–98
Sohn IG (1958) Chemical constituents of ostracods; some applications to paleontology and paleoecology. J Paleontol 32:730–736
Tappan H, Loeblich AR (1988) Foraminiferal evolution, diversification, and extinction. J Paleontol 62:695–714
Thomsen E, Rasmusen TL (2008) Coccolith-agglutinated foraminifera from the early Cretaceous and how they constructed their tests. J Foramin Res 38:193–214
Tibert NE, Scott DB (1999) Ostracodes and agglutinated Foraminifera as indicators of paleoenvironmental change in early Carboniferous brackish bay, Atlantic Canada. Palaios 14:246–260
Turpen JB, Angell RW (1971) Aspects of molting and calcification in the ostracods *Heterocypris*. Biol Bull 140:331–338
Van der Meeren T, Ito E, Verschuren D, Almendinger JE, Martens K (2011) Valve chemistry of *Limnocythere inopinata* (Ostracoda) in a cold arid environment—implications for paleolimnological interpretation. Palaeogeogr Palaeoclimatol Palaeoecol 306:116–126
Vannier J, Abe K (1992) Recent and early Palaeozoic Myodocope ostracodes: functional morphology, phylogeny, distribution and lifestyles. Palaeontology 35:485–517
Vannier J, Abe K (1993) Functional morphology and behaviour of *Vargula hilgendorfii* (Ostracoda: Myodocopida) from Japan, and discussion of its crustacean ectoparasites: preliminary results from video recordings. J Crustac Biol 13:51–76
Vannier J, Abe K, Ikuta K (1998) Feeding in myodocopid ostracods: functional morphology and laboratory observations from videos. Mar Biol 132:391–408
Vannier J, Thiéry A, Racheboeuf PR (2003) Spinicaudatans and ostracods (Crustacea) from the Montceau Lagerstätte (Late Carboniferous, France): morphology and palaeoenvironmental significance. Palaeontology 46:999–1030
van Raden UJ, Groeneveld J, Raitzsch M, Kucera M (2011) Mg/Ca in the planktonic foraminifera *Globorotalia inflate* and *Globigerinoides bulloides* from western Mediterranean plankton tow and core top samples. Mar Micropaleontol 78:101–112
Vogel N, Uthicke S (2012) Calcification and photobiology in symbiont-bearing benthic foraminifera and responses to a high CO_2 environment. J Exp Mar Biol Ecol 424–425:15–24
Weismann A (1887) On the signification of the polar globules. Nature 36:607–609
Wightman WG, Scott DB, Medioli FS, Gibling MR (1994) Agglutinated Foraminifera and thecoamoebians from the Late Carboniferous Sydney coalfield, Nova Scotia: paleoecology, paleoenvironments and paleogeographical implications. Palaeogeogr Palaeoclimatol Palaeoecol 106:187–202

Williams M, Leng MJ, Stephenson MH, Andrews JE, Wilkinson IP, Siveter DJ, Horne DJ, Vannier JMC (2006) Evidence that early Carboniferous ostracods colonised coastal flood plain brackish water environments. Palaeogeogr Palaeoclimatol Palaeoecol 230:299–318

Williams MW, Siveter DJ, Salas MJ, Vannier J, Popov LE, Pour MG (2008) The earliest ostracods: the geological evidence. Senckenb Lethaea 88:11–21

Yamada S, Keyser D (2010) Calcification of the marginal infold in podocopid ostracods. Hydrobiologia 638:213–222

Yamada S, Tsukagoshi A, Ikeya N (2005) Carapace formation of the podocopid ostracode *Semicytherura* species (Crustacea: Ostracoda). Lethaia 38:323–332

Zeebe RE, Sanyal A (2002) Comparison of two potential strategies of planktonic foraminifera for house building: Mg^{2+} or H^+ removal? Geochim Cosmochim Acta 66:1159–1169

Zeebe RE, Wolf-Gladrow D (2001) CO_2 in seawater: equilibrium, kinetics, isotopes: equilibrium, kinetics, isotopes. Elsevier Oceanography Series 65. Elsevier, Amsterdam, pp 1–360

Microbiological Survey in Two Arctic Fjords: Total Bacterial Number and Biomass Comparison of Hornsund and Kongsfjorden

Agnieszka Kalinowska, Anetta Ameryk and Katarzyna Jankowska

Abstract Two microbiological parameters: total bacterial number (TBN) and biomass (BBM) were studied in two Arctic fjords: Hornsund and Kongsfjorden. Samples were collected from three sampling points in each fjord, from various water depth layers: from the surface to 75 m depth. Total bacterial number and biomass were examined using the DAPI staining and direct count method. The greater amount of bacteria, as well as highest bacterial biomass were observed in the colder fjord Hornsund, located on the southern part of Spitsbergen. Local TBN maximum was found at a depth of 15 m at station KG2, which corresponds to the presence of pycnocline. Better adaptation of Arctic bacteria to the low water temperatures and availability of substrates coming from the bird colonies may explain the higher TBN and BBM in Hornsund. Moreover, greater phytoplankton abundance in Hornsund may have an impact on better feeding conditions. Detected increased amounts of particulate inorganic matter also provides favourable habitat for bacterial consortia.

Keywords Arctic · Hornsund · Kongsfjorden · Total bacterial number · Bacterial biomass

A. Kalinowska (✉) · K. Jankowska
Civil and Environmental Engineering Faculty, Gdańsk University of Technology, ul. Narutowicza 11/12, 80-952 Gdańsk, Poland
e-mail: agnieszka.kalinowska@poczta.onet.pl

A. Ameryk
Department of Fisheries, Oceanography and Marine Ecology, National Marine Fisheries Research Institute, ul. Kołłątaja 1, 81-332 Gdynia, Poland

T. Zielinski et al. (eds.), *Impact of Climate Changes on Marine Environments*, GeoPlanet: Earth and Planetary Sciences, DOI 10.1007/978-3-319-14283-8_9

1 Introduction

Bacteria play a vital role in every ecosystem. They reflect rapidly the changes in environment due to their short multiplication time. Because bacterial cells surface is relatively large compared to their volume, microorganisms produce significant amounts of carbon dioxide by their respiration and therefore influence CO_2 concentrations in the hydrosphere and, indirectly, in the atmosphere. Together with phytoplankton, microorganisms are responsible for the carbon cycle in freshwater and marine ecosystems (Hoppe et al. 2002). They are supporting not only the organic matter decay and mineralization, but also contribute to feeding the higher levels of trophic chains (so called microbial loop) (Azam et al. 1983; Pomeroy 1974) by incorporating the dissolved organic matter (DOM) into the trophic web. This activity of recycling the nutrients is recognized as the basic process providing proper Arctic marine ecosystems functioning (Nielsen and Hansen 1999; Iversen and Seuthe 2011; Lara et al. 2013). DOM of the land origin is used by microorganisms as the energy source in coastal and estuary areas (Zweifel and Hagstrom 1995; Moran and Hodson 1994). The main sources of bacteria nourishment are substances produced during organic matter and feces decay (Lampert 1978; Elbrächter 1991; Nagata and Kirchman 1992), as well as leftovers of zooplankton sloppy feeding and substances secreted by phytoplankton (Sharp 1977; Obernosterer and Herndl 1995). Availability of substrates and water temperature influence bacterial growth rates, when bacteriovorus and bacteriophages regulate the bacteria number and their biomass (Thingstad and Sakshaug 1990; Payet and Suttle 2008). Bacteria may live floating in the water as a single cells or in consortia on some solid particles; therefore the presence of particulate inorganic matter (PIM) creates a favourable habitat for larger colonies. Typical values of TBN in polar marine ecosystems usually vary between several thousands up to a few millions of cells in 1 cm^3 (Table 1).

Research in the Arctic region were limited in the past because of the access difficulties, but in recent decades polar region gained growing concern. Fjords of western Spitsbergen are one of the best explored arctic marine ecosystems, however microbial part of their biocenosis is still not so well known. In the past, bacterial abundance and biomass were investigated mostly in Kongsfjorden (Iversen and Seuthe 2011; Wängberg et al. 2008; Jankowska et al. 2005; Jiang et al. 2005). Also other detailed microbiological research were done in Kongsfjorden: isolation of cultivable bacteria (Prasad et al. 2014) and studying of microbial communities composition by the molecular methods (Piquet et al. 2010; Zeng et al. 2013). Hornsund is a less explored region—in 1985 the microbiological study was initiated by Zajączkowska and Zajączkowski (1989). Recently, single colony isolates from sediments were done by Rasol et al. (2014).

In this study the research of basic microbiological parameters was done to compare marine ecosystems of Hornsund and Kongsfjorden during the same summer. The purpose was to study the abundances, biomass and distribution of bacterial assemblages in the area of two arctic fjords. Also the relations with the physical water parameters was analyzed.

Table 1 Comparison of bacterial abundances reported in the Arctic region

Habitat, area	TBN [$\times 10^5$ cells cm^{-3}]	Method	Biomass	References
Fjordic waters, Hornsund, Spitsbergen	2.37–3.63	EPM, DAPI	8.3–16.1 mg C m^{-3}	This survey
Fjordic waters, Kongsfjorden, Spitsbergen	1.16–4.26	EPM, DAPI	1.4–12.4 mg C m^{-3}	This survey
Western Canada Basin	0.17–8.38 (mean 1.70)	FC, SYBR Green I	413.3 mg C m^{-3}	He et al. (2012)
Northern Barents Sea	3.6 (SD ± 3.0)	FC, SYBR Green I	7.1 (SD ± 6.1) mg C m^{-3}	Sturluson et al. (2008)
Across the Arctic Ocean	0.43–4.7	EPM, AODC, SYBR Green I	0.27–0.85 × 10^3 mg C m^{-2}	Steward et al. (2007)
Greenland and Norwegian Sea	0.22–6.01	EPM, AODC	0.5–15.0 mg C m^{-3}	Naganuma et al. (2006)
Fjordic water, Kongsfjorden, Spitsbergen	2.29–3.52	EPM, DAPI	3.7–8.01 mg C m^{-3}	Jiang et al. (2005)
Fjordic water, Kongsfjorden, Krossfjorden, Spitsbergen	7.8–80.0	EPM, AODC	2.5 × 10^3–14.8 × 10^3 mg C m^{-2}	Jankowska et al. (2005)
Barents Sea	4.1–41	EPM, DAPI	6.7–26.1 mg C m^{-3}	Howard-Jones et al. (2002)
Greenland Sea	0.97–28	EPM, DAPI	–	Børsheim (2000)
Across the Arctic Ocean	3.8–11.7	EPM, DAPI	0.28–0.77 × 10^3 mg C m^{-2}	Sherr et al. (1997)
Sea ice, Barents Sea and Laptev Sea	0.4–36.7	EPM, AODC	19.2–79.2 mg C m^{-2}	Gradinger and Zhang (1997)
Fjordic water, Hornsund, Spitsbergen	0.8–1.3	LM, stained erythrosine	–	Zajączkowska and Zajączkowski (1989)

EPM, DAPI—Epifluorescence microscope, stained DAPI
EPM, AODC—Epifluorescence microscope, stained acridine orange
FC, SYBR Green I—Flow cytometry, stained SYBR Green I
LM, stained erythrosine—Light microscope, stained erythrosine

2 Materials and Methods

2.1 Investigated Area

A microbiological study described in this paper was conducted on Spitsbergen (Svalbard) between the 27th of July and the 9th of August 2013, in the framework of the Growth of Arctic Marine Ecosystem project (GAME) from onboard the r/v Oceania (Institute of Oceanology, Polish Academy of Sciences). A research survey was conducted in two Arctic glacial fjords: Kongsfjorden and Hornsund. They are localized on the West Spitsbergen coast on various latitudes (Fig. 1). Both are influenced by relatively warm Atlantic waters (of summer temperatures around 4–6 °C) (Piwosz et al. 2009; Svendsen et al. 2002; Hop et al. 2002), however Hornsund (77°N, 16°E), in opposition to Kongsfjorden (79°N, 12°E), is more affected by the cold (below 0 °C) Arctic water masses coming from the Barents Sea on the eastern coast. Hornsund is almost two times shallower than Kongsfjorden (Tęgowski et al. 2008; Howe et al. 2003) and has more diversified bottom morphology—irregular shallower and deeper parts are located alternately along the fjord axis (Kowalewski et al. 1991). That may facilitate more intensive inflow, mixing and influence of Atlantic waters in Kongsfjorden rather than in Hornsund.

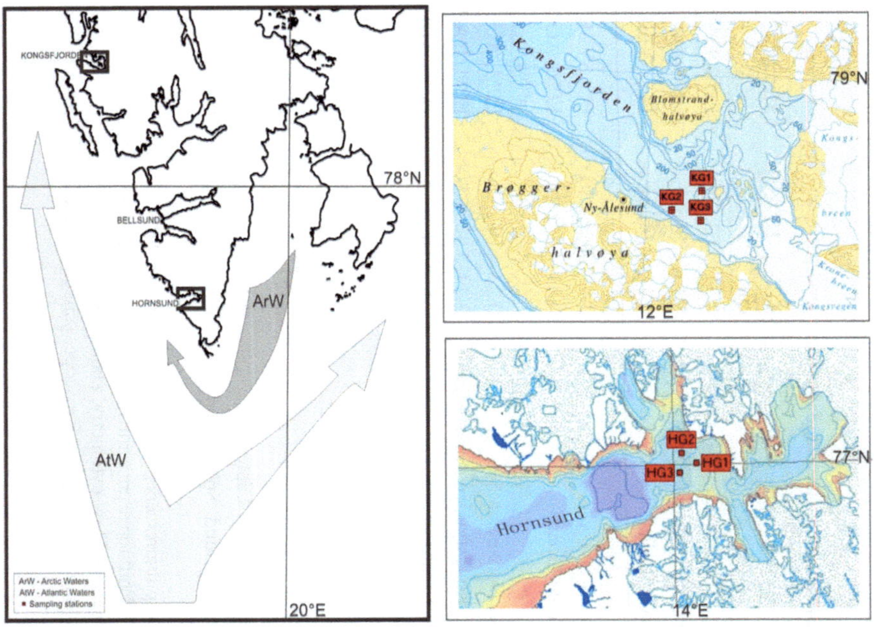

Fig. 1 Investigation area: West Spitsbergen and localization of two examined fjords. Sampling points labelled as *HG1, HG2, HG3* for Hornsund and *KG1, KG2, KG3* in Kongsfjorden are marked as *red rectangles*. Influence of the various water masses is shown by the *grey arrows* (maps thanks to: http://www.iopan.gda.pl/projects/Game/blog-pl.html)

Stratification of waters in the Arctic fjords occurs due to the different water masses inflows and influence of the atmospheric conditions. Several layers may be usually observed (Węsławski et al. 2006; maps at: www.iopan.gda.pl). Fjord surface waters are characterized by the lowest salinity (28–30 salinity units) because of glacier ice melting processes and freshwater inflow. They occupy the upper few meters of the water column but the additional topmost water layer may be separated and defined due to the strongest solar heating impact. In summer brackish top waters may reach even around 4 °C, which is higher than cold melted-ice waters below (Piwosz et al. 2009). Deeper strata under the surface waters usually reflect the activity of the ocean currents. Finally, bottom water masses are the coldest (temperature as low as −1.4 °C) and of the highest salinity in the entire water column (Svendsen et al. 2002).

2.2 Sampling

The research in each fjord was conducted on three sampling stations located on the vertexes of the equilateral triangle (arm of length around 1.5 km), in the inner part of the fjords. Water samples were taken from 4 depths in Hornsund (shallower fjord) and 5 depths from Kongsfjorden. Surface and fluorescence maximum were taken into consideration while establishing the sampling depths. At all stations the salinity and temperature of water were measured using CTD probe SEABIRD SBE 49. Water was sampled using Niskin bottles and stored in sterile plastic containers (50 ml) and fixed with particle-free formaldehyde solution (final concentration 2 %). Samples were kept at 4 °C until the microscopic analysis.

2.3 DAPI Staining and Direct Microscopic Counting

Total bacterial number and biomass were determined using DAPI (4′, 6-diamidino-2-phenylindole) direct count method (Porter and Feig 1980). 5 ml subsamples of each seawater sample were stained with DAPI solution to a final concentration of 1 μg/ml for 10–15 min and filtered on 25 mm polycarbonate filters (Milipore, 0.2 μm pore diameter).

Cells were counted using an epifluorescence microscope, the Nikon Eclipse 80i under 1,000-fold magnification. An HBO-103 W high pressure mercury burner, 330–380 nm excitation filter, 420 nm barrier filter and 400 nm dichroic mirror were used. The image analysis system of Świątecki (1997) was applied. It included the use of a PC, a high resolution CCD digital camera Nikon DS-5 Mc-U2 and MultiScan v.14.02 counting program with the modification of Świątecki (1997). Bacteria abundance, morphology and biomass was determined on the basis of 2 series of 10 fields from each depth. The mean value of two series was calculated for each sample. Blue fluorescing bacterial cells were classified into five size classes and three morphological types (cocci, rods and vibrio). Total bacterial number and biomass was calculated in accordance with Norland (1993). To verify the hypothesis of differences between parameters of two fjords, the t-Student test was applied.

3 Results

Salinity ranged from 30.1 to 34.7 in Hornsund and from 31.0 to 34.8 in Kongsfjorden. Fjord water temperature varied from 2.4 to 5.4 °C in Kongsfjorden (mean 3.6 °C) and from 1.6 to 5.4 °C in Hornsund (mean 2.9 °C) (Fig. 2).

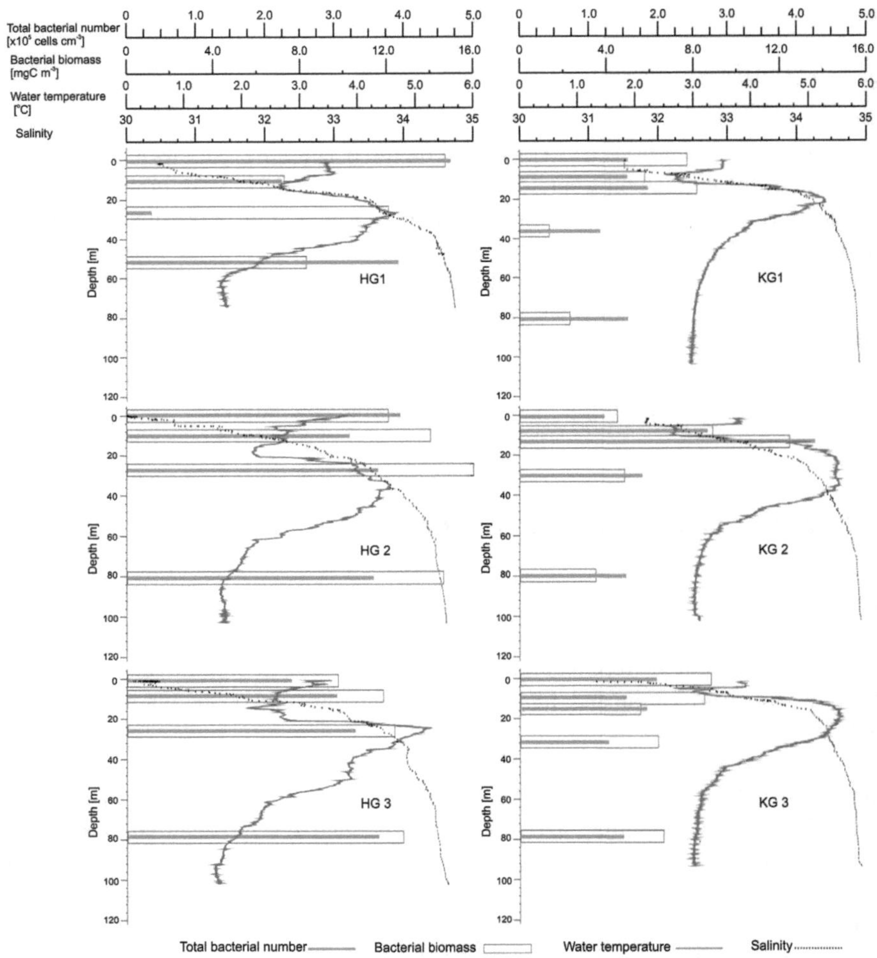

Fig. 2 Total bacterial number (TBN), biomass (BBM), water temperature and salinity as a function of depth at sampling stations

From all the samples the highest value of TBN was represented by the depth of 15 m on KG2 station, located adjacent to the fjord shore (4.3×10^5 cells cm^{-3}). The lowest value corresponded to 35 m on KG1 station (1.2×10^5 cells cm^{-3}). In Kongsfjorden, the increase in TBN was observed on 10 m ($1.5–2.7 \times 10^5$ cells cm^{-3}; mean 1.9×10^5 cells cm^{-3}) and 15 m ($1.8–4.3 \times 10^5$ cells cm^{-3}, mean 2.6×10^5 cells cm^{-3}). These are the depths corresponding to the change in temperature and salinity profile (water density change region), where suspended organic matter, being the food source for bacteria, tends to accumulate. Also high TBN and BBM values were recorded in surface water samples from Hornsund (Fig. 2). Regarding the number of bacteria in the entire water column, the highest value was represented by HG2 station (3.6×10^5 cells cm^{-3}) and the lowest was observed on the KG1 station (1.5×10^5 cells cm^{-3}).

BBM fluctuations were inconsistent with TBN fluctuations in case of depths at certain stations. The lowest BBM value were observed at 35 m on KG1 station (1.4 mg C m^{-3}) and the highest at 25 m on HG2 (16.1 mg C m^{-3}). However, the tendencies of entire water columns were similar: the highest BBM was recorded on HG2 (14.2 mg C m^{-3}) and the lowest on KG1 (5.1 mg C m^{-3}) (Fig. 2).

In general, total bacterial number and biomass were greater in Hornsund (mean for three stations: TBN = 3.3×10^5 cells cm^{-3}; BBM = 12.2 mg C m^{-3}) compared to the Kongsfjorden (TBN = 1.8×10^5 cells cm^{-3}; BBM = 6.3 mg C m^{-3}). Despite the lower mean water temperature, TBN and BBM was almost two times higher in Hornsund than in Kongsfjorden (Fig. 3a, b, d).

There was a significant difference in the salinity, temperature, TBN and BBM between fjords (Student's t-test; $p < 0.05$).

4 Discussion

The results presented in this paper generally correspond to the values from other publications from Arctic region. Noted differences in TBN and BBM counts may be the result of applying different counting methods and various stains (Table 1). DAPI, used in this research, is nowadays commonly used technique of direct counting under fluorescence microscopy. Acridine orange (AO) used widely in the past research (Jankowska et al. 2005; Naganuma et al. 2006 or Gradinger and Zhang 1997), caused overstating the results, which was mentioned in several publications (Posch et al. 2001; Sherr et al. 2001; Suzuki et al. 1993; Porter and Feig 1980).

Observed in this research water temperature values fall within the range of usual temperature of Arctic fjord waters, which may be as low as around −2 °C (Cottier et al. 2005; Lydersen et al. 2002; Jankowska et al. 2005) in the deepest parts of a fjord and as high as 4–6 °C (Piwosz et al. 2009) in the surface waters in summer. Water masses on the surface were warmer due to the constant solar radiation during polar day and several meters underneath the temperature decreased due to the glacier activity. In the layers localized beneath, the warming activity of Atlantic

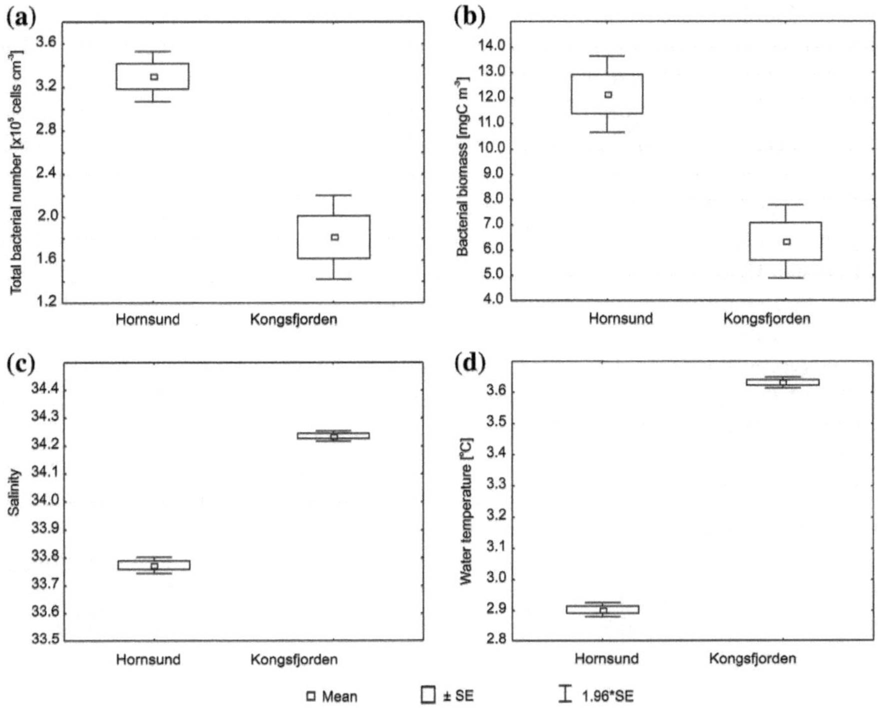

Fig. 3 Comparison between fjords **a** total bacterial number (TBN), **b** bacterial biomass (BBM), **c** salinity, **d** temperature

waters was noted. Similar hydrological conditions were observed by Iversen and Seuthe (2011) in Kongsfjorden in July 2006.

As Hornsund is more influenced by colder Arctic water masses and Kongsfjorden is under the strong inflow of warmer Atlantic currents (Beszczyńska-Möller et al. 1997; Svendsen et al. 2002, Fig. 1), the difference of the water temperature and salinity between fjords (Fig. 3c, d) is not surprising (Børsheim and Drinkwater 2014). Because the temperature of Kongsfjorden waters was higher, comparing to Hornsund, it might affect the bacterial abundance, however the exact dependence is not known.

In general, the Arrhenius equation states that a 10° increase in temperature doubles the chemical reaction rate. In case of microorganisms, the positive influence of temperature on bacteria (up to certain value) was studied by Li and Harrison (2001), Shiah et al. (2000) and Ameryk et al. (2005). However, in case of this research this tendency was not fulfilled—in Kongsfjorden where warmer water temperature was observed, the bacterial abundance and biomass were lower than in cold water Hornsund (Fig. 3). There are few possible explanations of such phenomenon. Bacteria adaptation to low temperature, food availability, predators pressure and bacteriophages action may be the parameters which probably act simultaneously.

As colder Hornsund, showing higher TBN and BBM values, is influenced by Arctic inflows, the bacteria present in this fjord may be also of Arctic origin, and therefore be better adapted to the lower water temperatures (Børsheim and Drinkwater 2014) than microorganisms occurring in more “Atlantic” Kongsfjorden.

Another parameter influencing bacterial abundance is food availability. According to Węsławski et al. (2006), bird populations in Hornsund are twice as numerous as in Kongsfjorden. Some species are even more abundant. For example, the little auk (*Alle alle*) colonies are consist of approximately 5 times more individuals in Hornsund than in Kongsfjorden. That corresponds to the greater consumption, and therefore easily affects the amount of faeces and organic matter which may be introduced to the fjord waters by e.g. surface runoff, providing an additional nourishment source for bacteria. Moreover, an increased phytoplankton biomass was observed in Hornsund (Piwosz et al. 2009). Phytoplankton itself, or its excreted products are important nutrient sources for bacteria (Sharp 1977; Obernosterer and Herndl 1995; Larsson and Hagsröm 1979).

The higher TBN detected in the surface waters of Hornsund may be positively correlated with the greater amounts of particulate organic matter (POM) and particulate inorganic matter (PIM) detected in this layer (database of IOPAN: GAME project, unpublished data). POM serves as the food source for microorganisms, and PIM coming e.g. from melting glaciers (Björkman et al. 2014) provides habitat for bacterial consortia, therefore favourable conditions were provided. Considering Kongsfjorden, higher TBN and BBM values in KG2 water column may be the result of activity of glacial river occurring in that region.

Water temperature profiles together with the salinity (Fig. 3c, d) enable us to localize the pycnocline, which as a result of water density changes, facilitates slowdown of microorganisms sedimentation rates and thus increased amounts of bacterial cells in this layer may be observed. Increased TBN values were observed in Kongsfjorden in proximity of the pycnocline, which was localized based on the salinity and temperature curves at depth of around 10–30 m. Similar phenomenon was registered in research made by Jankowska et al. (2005) in the same fjord.

Apart from factors mentioned above, other parameters, such as bacteriovory and lysis reduce bacterial number (Payet and Suttle 2008) which may play a significant role in Kongsfjorden, however the exact relationships were not studied in this research.

In the literature there are several informations regarding the biomass in Arctic waters, however different methods of obtaining those data (application of various conversion factors for carbon or calculating BBM from TBN) create significant problems. In the papers mentioned in Table 1, various units of biomass occur, therefore a straightforward comparison is impossible.

Acknowledgments The authors would like to thank the staff of the Oceania from AREX 2013 cruise for their cooperation during measurements and for making unpublished environmental data available. We would like to express special thanks to Jakub Kowalczyk, Daniel Rak and Jan Marcin Węsławski for help and support and IOPAS Archive of hydrographical data for providing necessary data.

References

Ameryk A, Podgórska B, Witek Z (2005) Dependance between bacterial production and environmental conditions in the Gulf of Gdańsk. Oceanologia 47(1):27–45

Azam F, Fenchel T, Field JG, Gray JS, Meyer-Reil LA, Thingstad F (1983) The ecological role of water-column microbes in the sea. Mar Ecol Prog Ser 10:257–263

Beszczyńska-Möller A, Węsławski JM, Walczowski W, Zajączkowski M (1997) Estimation of glacial meltwater discharge into Svalbard coastal water. Oceanologia 39(3):289–297

Björkman M, Zarsky J, Kühnel R, Hodson A, Sattler B, Psenner R (2014) Microbial cell retention in a melting high arctic Snowpack, Svalbard. Arct Antarct Alp Res 46(2):471–482

Børsheim KY (2000) Bacterial production rates and concentrations of organic carbon at the end of the growing season in the Greenland Sea. Aquat Microb Ecol 21:115–123

Børsheim KY, Drinkwater KF (2014) Differrent temperature adaptation in Arctic and Atlantic heterotrophic bacteria in the Barents Sea Polar Front region. J Mar Syst 130:160–166

Cottier F, Tverberg V, Inall M, Svendsen H, Nilsen F, Griffiths F (2005) Water mass modification in an Arctic fjord through cross-shelf exchange: the seasonal hydrography of Kongsfjorden Svalbard. J Geophys Res 110(C12):C12005

Elbrächter M (1991) Faeces production by dinoflagellates and other small flagellates. Mar Microb Food Webs 5:189–204

Gradinger R, Zhang Q (1997) Vertical distribution of bacteria in Arctic Sea ice from the Barents and Laptev Sea. Polar Biol 17:448–454

He J, Zhang F, Lin L, Ma Y, Chen J (2012) Bacterioplankton and picoplankton abundance, biomass and distribution in the Western Canada Basin during summer 2008. Deep-Sea Res II 81–84:36–45

Hop H, Pearson T, Hegseth E, Kovacs KM, Wiencke C, Kwaśniewski S, Eiane K, Mehlum F, Gulliksen B, Włodraska-Kowalczuk M, Lydersen C, Węsławski JM, Cochrane S, Gabrielsen GW, Leakey RJG, Lønne OJ, Zajączkowski M, Falk-Petersen S, Kendall M, Wängberg S, Bishof K, Voronkov AY, Kovaltchouk NA, Wiktor J, Poltermann M, di Prisco G, Papucci C, Gerland S (2002) The marine ecosystem of Kongsfjorden, Svalbard. Polar Res 21(1):167–208

Hoppe H, Gocke K, Koppe R, Begler C (2002) Bacterial growth and primary production along a north-south transect of the Atlantic Ocean. Nature 416:168–171

Howard-Jones M, Ballard VD, Allen AE, Frischer ME, Verity PG (2002) Distribution of bacterial biomass and activity in the marginal ice zone of the central Barents Sea during summer. J Mar Syst 38(1–2):77–91

Howe JA, Moreton SG, Morri C, Morris P (2003) Multibeam bathymetry and the depositional environments of Kongsfjorden and Krossfjorden, western Spitsbergen, Svalbard. Polar Res 22 (2):301–316

Iversen RK, Seuthe L (2011) Seasonal microbial processes in a high-latitude fjord (Kongsfjorden, Svalbard): I. Heterotrophic bacteria, picoplankton and nanoflagellates. Polar Biol 34(5):731–749

Jankowska K, Włodarska-Kowalczuk M, Wieczorek P (2005) Abundance and biomass of bacteria in two Arctic glacial fjords. Polish Polar Res 26(1):77–84

Jiang X, He J, Cai M (2005) Abundance and biomass of heterotrophic microbes in the Kongsfjorden, Svalbard. Acta Oceanol Sinica 24:143–152

Kowalewski W, Rudowski S, Zalewski SM (1991) Seismoacoustic studies in Hornsund, Spitsbergen. Pol Polar Res 12:353–361

Lampert W (1978) Release of dissolved grazing zooplankton carbon by grazing zooplankton. Limnol Oceanogr 23(4):831–834

Lara E, Arrieta JM, Garcia-Zaradona I, Boras JA, Duarte CM, Agustí S, Wassmann PF, Vaqué D (2013) Experimental evaluation of the warming effect on viral, bacterial and protistan communities in two contrasting Arctic systems. Aquat Microb Ecol 70:17–32

Larsson U, Hagström A (1979) Phytoplankton exudate release as an energy source for the growth of pelagic bacteria. Mar Biol 52:199–206

Li WKW, Harrison WG (2001) Chlorophyll, bacteria and picophytoplankton in ecological provinces of the North Atlantic. Deep-Sea Res II 48:2271–2293

Lydersen C, Nøst OA, Lovell P, McConnell BJ, Gammelsrød T, Hunter C, Fedak MA, Kovacs KM (2002) Salinity and temperature structure of a freezing Arctic fjord—monitored by white whales (Delphinapterus leucas). Geophys Res Lett 29(23):2119–2123

Moran MA, Hodson RE (1994) Dissolved humic substances of vascular plant origin in a coastal marine environment. Limnol Oceanogr 39(4):762–771

Naganuma T, Kimura H, Karimoto R, Primenov NV (2006) Abundance of planctonic thraustochytrids and bacteria and the concentration of the particulate ATP in the Greenland and Norwegian Sees. Polar Biosci 20:37–45

Nagata T, Kirchman DL (1992) Release of dissolved organic matter by heterotrophic protozoa: implications for microbial foodwebs. Archiv für Hydrobiologie Beiheft Ergebnisse Limnologie 35:99–109

Nielsen TG, Hansen BW (1999) Plankton community structure and carbon cycling on the western coast of Greenland during the stratified summer situation, I. Hydrography, phytoplankton and bacterioplancton. Aquat Microb Ecol 16:205–216

Norland S (1993) The relationship between biomass and volume of bacteria. In: Kemp PF, Sherr BF, Sherr EB, Cole JJ (eds) Handbook of methods in aquatic microbial ecology. Lewis Publishers, New York, pp 303–308

Obernosterer I, Herndl G (1995) Phytoplankton extracellular release and bacterial growth: dependence on the inorganic N: P ratio. Mar Ecol Prog Ser 116:247–257

Payet JP, Suttle CA (2008) Physical and biological correlates of virus dynamics in the southern Beaufort Sea and Amundsen Gulf. J Mar Syst 74:933–945

Piquet AM-T, Scheepens JF, Bolhuis H, Wiencke C, Buma AGJ (2010) Variability of protistan and bacterial communities in two Arctic fjords (Spitsbergen). Polar Biol 33:1521–1536

Piwosz K, Walkusz W, Hapter R, Wieczorek P, Hop H, Wiktor J (2009) Comparison of productivity and phytoplankton in a warm (Kongsfjorden) and a cold (Hornsund) Spitsbergen fjord in mid-summer 2002. Polar Biol 32(4):549–559

Pomeroy LR (1974) The ocean's food web, a changing paradigm. Bioscience 24(9):499–504

Porter KG, Feig YS (1980) The use of DAPI for identifying and counting aquatic microflora. Limnol Oceanogr 25(5):943–948

Posch T, Loferer-Krößbacher M, Gao G, Alfreider A, Pernhalter J, Psenner R (2001) Precision of bacterioplankton biomass determination: a comparison of two fluorescent dyes, and of allometric and linear volume-to-carbon conversion factors. Aquat Microb Ecol 25:55–63

Prasad S, Manasa P, Buddhi S, Tirunagari P, Begum Z, Rajan S, Shivaji S (2014) Diversity and bioprospective potential (cold-active enzymes) of cultivable marine bacteria from the subarctic glacial fjord, Kongsfjorden. Curr Microbiol 68:233–238

Rasol R, Rashidah AR, Siti Nur Nazucha R, Smykla J, Wan Maznach WO, Alias SA (2014) Psychrotropic Lipase producers from arctic soil and sediment samples. Pol J Microbiol 63 (1):75–82

Sharp JH (1977) Excretion of organic matter by marine phytoplankton: do healthy cells do it? Limnol Oceanogr 22(3):381–399

Sherr B, Sherr E, del Giorgio P (2001) Enumeration of total and Highly Active Bacteria. Methods Microbiol 30(8):129–159

Sherr E, Sherr B, Fessenden L (1997) Heterotrophic protist in the central Arctic Ocean. Deep-Sea Res II 44:1665–1682

Shiah FK, Gong GC, Chen TY, Chen CC (2000) Temperature dependence of bacterial specific growth rates on the continental shelf of the East China Sea and its potential application in estimating bacterial production. Aquat Microb Ecol 22:155–162

Steward GF, Fandino LB, Hollibaugh JT, Whitledge TE, Azam F (2007) Microbial biomass and viral infections of heterotrophic procaryotes in the sub-surface layer of the central Arctic Ocean. Deep-Sea Res I 54:1744–1757

Sturluson M, Gissel Nielsen T, Wassmann P (2008) Bacterial abundance, biomass and production during spring blooms in the northern Barents Sea. Deep-Sea Res II 55:2186–2198

Suzuki TM, Sherr EB, Sherr BF (1993) DAPI direct counting underestimates bacterial abundances and average cell size compared to AO direct counting. Limnol Oceanogr 38(7):1566–1570

Svendsen H, Beszczyńska-Möller A, Hagen JO, Lefauconnier B, Tverberg V, Gerland S, Ørbæk JB, Bischof K, Papucci C, Zajączkowski M, Azzolini R, Bruland O, Wiencke C, Winther J, Dallmann W (2002) The physical environment of Kongsfjorden—Krossfjorden, an Arctic fjord system in Svalbard. Polar Res 21(1):133–166

Świątecki A (1997) Zastosowanie wskaźników bakteriologicznych w ocenie wód powierzchniowych. (Application of bacteriological indicators in surficial water quality assessment). WSP Olsztyn, pp 1–105

Tęgowski J, Giżejewski J, Zieliński A (2008) Extraction of seabed geomorphologic features of svalbard fjords using high-resolution side scan sonar. Hydroacoustics 11:401–410

Thingstad TF, Sakshaug E (1990) Control of phytoplankton growth in nutrient recycling ecosystems: theory and terminology. Mar Ecol Prog Ser 63:261–271

Wängberg S-Å, Andreasson KIM, Gustavson K, Reinthaler T, Henrikson P (2008) UV-B effects on microplankton communities in Kongsfjord, Svalbard—A mesocosm experiment. J Exp Mar Biol Ecol 365(2):156–163

Węsławski JM, Kwaśniewski S, Stempniewicz L, Błachowiak-Samołyk K (2006) Biodiversity and energy transfer to top trophic levels in two contrasting Arctic fjords. Pol Polar Res 27(3):1–20

Zajączkowska B, Zajączkowski M (1989) Quantitative microbiological survey in Hornsund, SW Spitsbergen. Reconnaissance study in summer 1985. Bull Pol Acad Sci 37:79–84

Zeng Y-X, Zhang F, He J-F, Lee SH, Qiao Z-Y, Yu Y, Li H-R (2013) Bacterioplankton community structure in the Arctic waters as revealed by pyrosequencing of 16S rRNA genes. Antonie Van Leeuwenhoek 103:1309–1319

Zweifel UL, Hagstrom Å (1995) Total counts of marine bacteria include a large fraction of non-nucleoid-containing bacteria (ghosts). Appl Environ Microbiol 61(6):2180–2185

New Methods in the Reconstruction of Arctic Marine Palaeoenvironments

Magdalena Łącka, Joanna Pawłowska and Marek Zajączkowski

Abstract In recent years, numerous new proxies have been developed for the reconstruction of past environmental conditions in the polar regions. In this review we focus on the selected methods that are used in the reconstruction of the Arctic marine paleoenvironments, i.e., organic (IP_{25} and PIP_{25} index, U_{37}^{K} and $U_{37}^{K'}$ and GDGT palaeothermometry) and inorganic geochemical indices (Mg/Ca and fragmentation/dissolution analysis) as well as genetic (ancient DNA) and physical (XRF, magnetic susceptibility) proxies. A brief description of each of them is presented with example applications.

1 Introduction

The surface temperatures of the Earth are most reliably known for 1850 to the present (Jones et al. 2001; Brohan et al. 2006) because there has been reasonable global coverage of stations that systematically measured temperature for that period. The instrumental record is generally considered to not to be long enough to give a complete picture of climatic variability. Recent records are also likely influenced by human activity (Barnett et al. 1999). Therefore, it is crucial to extend the record of climatic changes beyond the era of instrumental measurements to understand how large natural climatic variability can be, how rapidly the climate may change, which internal mechanisms drive climatic changes on regional and global scales, and what external or internal forcing factors control them (Houghton et al. 1996).

The Arctic is generally recognised as an area where climate changes have a disproportionally large impact (ACIA 2005). The Arctic air temperature during the 20th century was the highest of the past 400 years (Serreze et al. 2000), and IPCC reports (2007, 2013) predict that this trend will continue in the coming decades.

M. Łącka (✉) · J. Pawłowska · M. Zajączkowski
Institute of Oceanology, Polish Academy of Sciences,
Powstańców Warszawy 55, 81-712 Sopot, Poland
e-mail: mlacka@iopan.gda.pl

T. Zielinski et al. (eds.), *Impact of Climate Changes on Marine Environments*,
GeoPlanet: Earth and Planetary Sciences, DOI 10.1007/978-3-319-14283-8_10

Recent global warming, which has had wide impacts in the Arctic regions (IPCC 2007), has been demonstrated as the cause of significant mass loss of the Greenland ice sheet (Chen et al. 2006), melting of the permafrost (ACIA 2005) and a progressive decrease of sea ice over the last 30 years (Comiso et al. 2008; Schiermeier 2012). Moreover, the increasing sea surface temperature may lead to migration of the species further north (e.g., Walther et al. 2002) and extreme weather events in mid-latitudes (Francis and Vavrus 2012). All of these rapid recent changes emphasise the need for information about past natural climatic variability.

Palaeoceanography is the study of the history of the oceans in the geologic past with regard to circulation, chemistry, biology, geology and patterns of sedimentation and biological productivity. Palaeoceanographic studies using environmental models and different proxies enable the scientific community to assess the role of oceanic processes in the global climate by the reconstruction of the past climate at various intervals. Palaeoceanography makes use of so-called proxy methods to infer information about the past state and evolution of the world's oceans. Only by understanding past climate variability we are able to determine the driving mechanisms of global climate change.

To understand recent and predicted climate fluctuations, it is essential to comprehend the natural amplitudes and rates of changes in the past, such as the extents of Arctic glaciers, Arctic Ocean sea ice, and fluctuations in sea water temperature. Commonly used proxy indicators in palaeoceanography include the assemblage composition of planktic and benthic microfossils, stable oxygen and carbon isotopes, chemical tracers, composition and accumulation rates of organic matter and the molecular composition of biomarkers. In this paper, we focus on selected recently developed methods that are used in the reconstruction of the Arctic marine paleoenvironments.

2 New Methods and Example Applications

2.1 Organic Geochemical Proxies

Progress in developing new techniques in analytical chemistry and applying organic geochemical indicators to palaeoceanography has been made over the last decade (Sachs et al. 2013). The source of organic geochemical proxies is organic matter produced principally by photosynthetic plants, bacteria, and archaea and incorporated into the marine sediments.

2.1.1 Sea Ice Biomarker IP_{25} and PIP_{25} Index

One of the novel biomarker proxies that have been discovered in Arctic environment is a mono-unsaturated C_{25} highly branched isoprenoid (HBI, which is a molecule with 25 carbon atoms and one double bond at $C_{23\text{–}24}$ carbon atoms; Belt et al. 2007). The sea ice-specific HBI called IP_{25} (Ice Proxy with 25 carbon atoms) is biosynthesised by sea ice diatoms (e.g., Belt et al. 2007) from the genus *Haslea* (Belt et al. 2007; Belt and Müller 2013; Stoynova et al. 2013; Xiao et al. 2013) and a species from the *Pleurosigma* genus (*Pleurosigma stuxbergii* var. *rhomboides*; Brown et al. 2014).

HBIs (including IP_{25}) are preserved in sediment and laboratory based experiments showed their good resistance to transformations by photo-oxidation (in the upper water layer), biodegradation (at the sediment-water interface; Belt et al. 2000; Robson and Rowland 1986; Rontani et al. 2014), which makes it useful as a proxy for sediments from the Holocene and as old as 2.2 Ma (Stein and Fahl 2012). IP_{25} is very sensitive and can be detected from <1 g of sediment, which enables detailed and high resolution sediment core analyses with sub-decadal resolution (Belt et al. 2007).

The presence and variability of IP_{25} in down-core sediments from several Arctic regions, including the Barents Sea (Vare et al. 2010; Berben et al. 2014), Fram Strait (Müller et al. 2009), northern Norway (Cabedo-Sanz et al. 2013), northern Iceland (Massé et al. 2008; Andrews et al. 2009; Axford et al. 2011), the Canadian Arctic Archipelago (Vare et al. 2009; Belt et al. 2007, 2008, 2010; Brown et al 2011) and the eastern Arctic Ocean (Fahl and Stein 2012), has been interpreted as a direct indication of palaeo spring sea ice coverage and its change over time.

The distribution of IP_{25} in a sediment core correlates well with historic data of sea ice occurrence (Massé et al. 2008; Alonso-Garcia et al. 2013), other sea ice proxy data, such as IRD, bowhead whale remains (Vare et al. 2009) or dinoflagellates (Stein et al. 2014), and instrumental data (Müller et al. 2011), which supports the applicability of this biomarker. Moreover, there is an established relationship between IP_{25} abundances from surface sediments in different Arctic locations and known recent sea ice conditions derived from satellite data (Müller et al. 2011; Navarro-Rodriguez et al. 2013; Stoynova et al. 2013; Xiao et al. 2013).

However, Belt et al. (2007) and Müller et al. (2011) point out that the absence of IP_{25} in marine (Arctic) sediments might reflect either a lack of sea ice or indicate a permanent ice cover that prevents any algal growth. The additional use of the phytoplankton-derived biomarker brassicasterol (or dinosterol) as an indicator of open-water conditions facilitates the environmental reconstruction of ambiguous IP_{25} signals (Müller et al. 2009). Phytoplankton-IP_{25} indices ("PIP_{25}-Index") might be used for the semi-quantitative evaluation of palaeo sea ice conditions that can be incorporated into models for forecasting further climate change. The PIP_{25} index accounts for the (spring/summer) algal activity beneath the sea ice (mainly ice algae), at the ice-edge (ice and phytoplankton algae), and in ice-free areas (phytoplankton) and thus allows a rough estimate of the spatial and temporal extent of the sea ice cover. The absence of both biomarkers might demonstrate a permanent ice cover, whereas the absence of IP_{25} with elevated levels of brassicasterol suggests ice-free conditions. On the other hand, the occurrence of high (but variable) abundances of both biomarkers reflects the seasonal ice margin (Fig. 1; Belt and Müller 2013).

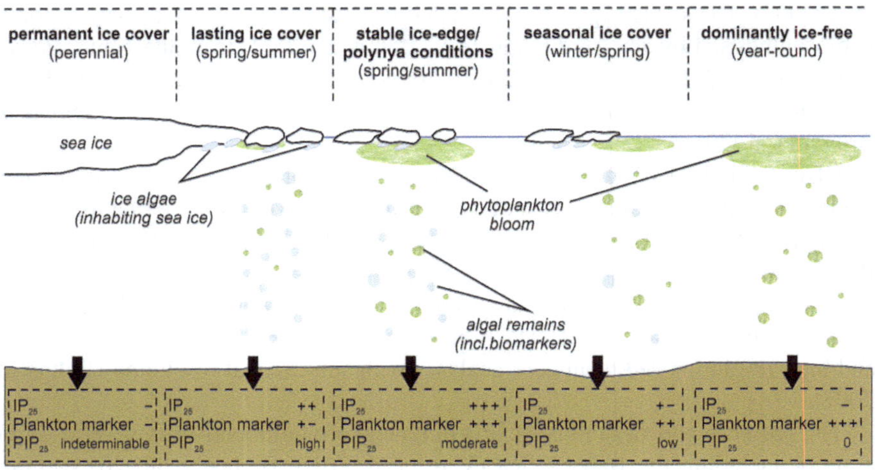

Fig. 1 Schematic illustration of the sedimentary contents of IP_{25} and the phytoplankton-derived biomarkers and resulting PIP_{25} indices for each setting (modified after Belt and Müller 2013)

2.1.2 Alkenone Unsaturation Indices U_{37}^{K} and $U_{37}^{K'}$

The index U_{37}^{K} (U = unsaturation, K = ketone, 37 = chain length of ketone) was first proposed by Brassell et al. (1986) to quantify the unsaturation of alkenones. Alkenones are derived from marine phytoplankton of the class Prymnesiophyceae; in cold, polar waters, they are limited to the coccolithophorid *Emiliania huxleyi* (Volkman et al. 1980; Bendle et al. 2005). This alga is widespread in the euphotic zone of all modern oceans, from tropical to polar waters (Okada and Honjo 1973; Okada and McIntyre 1979), and is tolerant of different nutrient and light conditions (Prahl et al. 2003). *E. huxleyi* responds to changes in water temperature by altering the molecular composition of their cell membranes; as the water temperature decreases, they increase the production of alkenones (Bradley 2014).

The standard U_{37}^{K} index measures the relative abundance of the di-, tri- and tetra-unsaturated C_{37} alkenones ($C_{37:2}$, $C_{37:3}$, $C_{37:4}$) in sediments and is defined as follows:

$$U_{37}^{K} = \frac{C_{37:2} - C_{37:4}}{C_{37:2} + C_{37:3} + C_{37:4}}$$

$C_{37:4}$ alkenone is rarely detected in open sea sediments from low- and mid-latitudes. Thus, the equation for those areas can be written as (Prahl and Wakeham 1987):

$$U_{37}^{K'} = \frac{C_{37:2}}{C_{37:2} + C_{37:3}}$$

$C_{37:4}$ alkenone becomes more important with decreasing temperature (<10 °C), so U_{37}^{K} is more reliable in high latitude areas, such as the northern North Atlantic and the Southern Oceans (Rosell-Melé et al. 1995; Sikes et al. 1997; Andresen et al. 2013).

Global surveys of surface sediments and water column particulate organic matter have revealed a strong correlation between U^K_{37} and sea surface temperature (SST, reviewed by Herbert 2003). Therefore, U^K_{37} can be converted into realistic estimates of the mean annual temperature at the sea surface using the regression from a core-top calibration in the northeastern Atlantic. The error of the regression is 0.5 °C (Rosell-Melé et al. 1995).

$$SST(^{\circ}\mathrm{C}) = \frac{U^K_{37} - 0.162}{0.029}$$

Alkenones are preserved in sediments even after the dissolution of the calcareous remains of their producers (Sachs et al. 2013); thus, they represent an important biomarker for the reconstruction of past sea surface temperatures (Brassell et al. 1986; Brassell 1993; Marlowe et al. 1984; Prahl and Wakeham 1987; Volkman et al. 1995). Moreover, alkenones appear to be relatively stable and are also present in older sedimentary rocks from the Cretaceous (Farrimond et al. 1987) and Eocene (Marlowe et al. 1984; Dzvonik 1996).

The limiting factor for U^K_{37} usage may be a solid ice cover that affects light penetration (Bendle and Rosell-Melé 2004) and thus the distribution of alkenones. The biochemical function of alkenones is still unknown, and U^K_{37} may be affected by changes in the species composition, seasonality and depth habitat changes of haptophytes, nutrient utilisation (Prahl et al. 2005) and oxic degradation (Hoefs et al. 1998; Gong and Hollander 1999). Moreover, in the case of all SST proxies, sediment resuspension (Bendle and Rosell-Melé 2004) and/or lateral advection (Andresen et al. 2013 and references therein) may be sources of errors.

Bendle and Rosell-Melé (2004) detected alkenones in all of the major water masses of the Nordic seas in an SST range of −0.5 to 13 °C and sea surface salinity (SSS) values of 29.6–35.6. Hence, U^K_{37} has been successfully used to the reconstruction of SST in those regions, e.g., in the Fram Strait (Rueda et al. 2013).

The relatively high abundances of $C_{34:4}$ alkenone in the glacial period indicate either transport and/or development of cold and low salinity water masses or increases in the input of fresh water from rivers (Seki et al. 2005). Therefore, the relative abundance of tetra-unsaturated alkenones ($C_{37:4}$ alkenones) could potentially be used as a qualitative indicator of palaeo-salinity (e.g., Rosell-Melé et al. 1998; Schulz et al. 2000; Harada et al. 2003).

2.1.3 GDGT Palaeothermometry

In recent years, several proxies that are based on the relative abundances of glycerol dialkyl glycerol tetraethers (GDGTs) have been developed to complement and extend the alkenone proxy. GDGTs are relatively large (up to 86 carbon atoms) membrane lipids produced by archaea and some bacteria. They are commonly found in a variety of environments (lakes, soils, oceans). There are two main types of GDGTs:

isoprenoidal GDGTs (isoGDGTs) and non-isoprenoidal branched GDGTs (brGDGTs). Isoprenoidal GDGTs are characteristic lipids of archaea and are mainly from the phylum Thaumarchaeota (formerly a division of Crenarchaeota; Brochier-Armanet et al. 2008; Spang et al. 2010). Thaumarchaeota and their GDGTs are commonly found in oceans, lakes and soils (Karner et al. 2001; Ochsenreiter et al. 2003; Leininger et al. 2006; Llirós et al. 2010; Schouten et al. 2012). Thaumarchaeota are the only archaea known to make the distinctive GDGT crenarchaeol, which is a diagnostic biomarker for these species (Sinninghe-Damste et al. 2002; Pearson et al. 2004; de la Torre et al. 2008). Branched GDGTs (brGDGTs) are diagnostic of bacteria (Weijers et al. 2006) and are commonly found in soils, peats, lakes, and marginal/deltaic environments but are not present in pelagic marine environments (Hopmans et al. 2004). The producers of most brGDGTs have not been identified; thus, the physiological relationships between the structure and abundance of brGDGTs to environmental factors are not understood. Weijers et al. (2007a, b) found that soil temperature and soil pH was statistically related to the brGDGT distribution. They proposed indices related to the degree of cyclisation and methylation of brGDGTs are called CBT (Cyclisation of Branched Tetraethers) and MBT (Methylation of Branched Tethraethers), respectively.

IsoGDGTs are characterised by the presence of cyclopentane moieties. In response to increasing temperature, the number of cyclopentane moieties of GDGT membrane lipids also increases (De Rosa et al. 1980; Uda et al. 2001; Wuchter et al. 2004; Schouten et al. 2007). Schouten et al. (2002) developed an index to mathematically represent this degree of cyclisation called TEX_{86} (the TetraEther indeX of lipids with 86 carbons). The TEX_{86} index was determined for a range of marine core top sediments and found to correlate well with mean SST (Schouten et al. 2002):

$$TEX_{86} = 0.015 * SST + 0.28\left(r^2 = 0.92,\ n = 43\right)$$

where:

$$TEX_{86} = \frac{[GDGT-2] + [GDGT-3] + [Cren']}{[GDGT-1] + [GDGT-2] + [GDGT-3] + [Cren']}$$

The numbers indicate the number of cyclopentane moieties, and Cren ′ stands for crenarchaeol regioisomer.

This empirical correlation was confirmed by experimental studies that showed that TEX_{86} is linearly correlated with temperature and is not influenced by salinity and nutrient availability (Wuchter et al. 2004) or grazing (Huguet et al. 2006). The diagenetic stability of the TEX_{86} proxy was also tested. Although isoGDGTs degrade faster than brGDGTs (Huguet et al. 2008), diagenetic degradation does not affect the TEX_{86} index within the analytical error (Sinninghe-Damste et al. 2002; Schouten et al. 2004; Kim et al. 2009; Huguet et al. 2009).

The current TEX_{86} calibration dataset for the marine realm includes more than 350 core tops from around the global ocean, and the relationship between TEX_{86}

and SST remains largely linear. However, in the polar oceans, the changes in TEX_{86} with temperature were relatively minor (Kim et al. 2008, 2010). Kim et al. (2010) proposed an alternate functional form of TEX_{86} called TEX_{86}^{L} (L represents low temperature) that has better predictability at low temperatures but is less precise given the entire global SST range:

$$TEX_{86}^{L} = \log\left(\frac{[GDGT-2]}{[GDGT-1]+[GDGT-2]+[GDGT-3]}\right)$$

Other alternative calibrations for polar regions have been proposed based on the disagreements with other temperature proxies ($pTEX_{86}$; Hollis et al. 2012), the influence of terrigenous input (TEX_{86}; Sluijs et al. 2006), and the influence of subsurface GDGTs in the sedimentary pool (Kim et al. 2012).

Sluijs et al. (2006) and Weijers et al. (2007a, b) applied the TEX_{86} and MBT/CBT proxies to the sedimentary sequence from the Lomonosov Ridge. They reconstructed surprisingly warm temperatures for the high Arctic that suggested a strongly reduced latitudinal gradient in temperature during the Palaeocene–Eocene Thermal Maximum Ho et al. (2014) reevaluated the applicability of TEX_{86} and TEX_{86}^{L} in polar and subpolar regions based on 160 surface sediment samples from the Arctic and North and South Pacific. They reported overestimated SST-derived TEX_{86}/TEX_{86}^{L} values at many Arctic sites and a robust relationship between TEX_{86}/TEX_{86}^{L} and SST in the Southern Ocean and the Pacific Subarctic Front zone. They concluded that the use of TEX_{86} with a global calibration is suitable for the Southern Ocean and the Pacific Subarctic Front zone. They also suggested the use of a regional TEX_{86}^{L} calibration in areas where the difference between the regional TEX_{86}^{L} calibration and SST was observed.

2.2 Ancient DNA

Ancient DNA (aDNA) research is defined broadly as the retrieval of DNA sequences from museum specimens, archaeological finds, fossil remains, and other unusual sources of DNA. The molecular cloning of DNA sequences of quagga (an extinct subspecies of the plains zebra; Higuchi et al. 1984) and an Egyptian mummy (Pääbo 1985) were the first successes in retrieving ancient DNA sequences. The study of ancient DNA has the allure of time travel and has attracted much attention and many practitioners. Most authenticated ancient DNA studies have analysed hard or soft tissue remains of flora and fauna from the late Pleistocene and Holocene. A further step towards palaeogenetics was the demonstration that sediments, as well as ice and permafrost, have the potential to preserve fossil DNA. Recent studies have shown that sedimentary genetic signals of bacteria, plant, and animal communities can be preserved for considerable periods of time in both permafrost and temperate conditions (Anderson-Carpenter et al. 2011; Pääbo et al. 2004; Willerslev et al. 2003, 2004a, 2007). For example, plant aDNA has been retrieved from 300–400 kyr-old permafrost sediments (Willerslev et al. 2003), and bacterial DNA sequences have been found in sediments that are more than half a million years old (Willerslev et al. 2004b).

With the advent of high-throughput sequencing (HTS) technologies, it is possible to thoroughly investigate the taxonomic composition of numerous and diverse sediment samples using environmental sequencing approaches (Bik et al. 2011; Pawlowski et al. 2011). Recent studies have shown that ancient plankton DNA can be recovered from Holocene marine sediments from species that do not leave fossils in sediment (Coolen et al. 2006, 2007, 2013; Boere et al. 2009, 2011a, b). The results of these studies have been promising and show that the marine sediments are an excellent DNA repository that can be used for the assessment of marine biodiversity. Several studies have used palaeogenetic data to trace planktonic successions during the Holocene (e.g., Boere et al. 2011b; Coolen et al. 2013) and to investigate the histories of dinoflagellates (Boere et al. 2009), haptophytes (Coolen et al. 2006), radiolarians and foraminifera (Lejzerowicz et al. 2013; Pawłowska et al. 2014).

Comparative multi-proxy surveys have been used to test the accuracy of ancient DNA approaches in reconstructing past planktonic communities in the Antarctic (Boere et al. 2009, 2011a) and benthic foraminiferal assemblages in the deep Atlantic (Lejzerowicz et al. 2013) and Arctic (Pawłowska et al. 2014). Of the approaches that have been used, ancient DNA records provide the most information. Molecular approaches have been proven to be useful in identifying predominant and potentially important taxa that were not revealed by microfossils or cysts. Despite the differences between the taxonomic compositions of microfossils (Lejzerowicz et al. 2013; Pawłowska et al. 2014), cysts (Boere et al. 2009), dinosterols (Boere et al. 2009, 2011a) and molecular approaches provided complementary information. However, it is important to use a palaeogenetic approach in combination with other independent methods to gain a better understanding of the palaeoenvironmental information that is inferred from other proxies.

2.3 Elemental Composition of Calcareous Tests: Mg/Ca Ratio

Most calcareous shells are composed of calcite and aragonite with the chemical formula $CaCO_3$. However, in almost all shells, especially those formed in shallow shelf waters, Mg might substitute for Ca during the formation of biogenic calcium carbonate. The temperature sensitivity of foraminiferal Mg/Ca ratios was first reported by Chave (1954) and is described by the empirical relationship:

$$\mathrm{T} = \mathrm{A} * 10^{(\mathrm{B}*[\mathrm{Mg/Ca}])}$$

where A and B are constants and are dependent on the species. Chave (1954) recognised that the Mg content in calcareous skeletons is the highest in the tropics and suggested that the substitution of Ca by Mg is facilitated by high temperatures.

Positive correlations between the Mg/Ca ratio and temperature have since been found in many kinds of organisms, including benthic foraminifera, ostracods, coccoliths and corals (Wefer et al. 1999; Barker et al. 2005 and references therein).

In cold, near surface environments at high latitudes, the Mg/Ca ratio is lower than in warmer waters, so the changes with temperatures are small (Meland et al. 2006). Therefore, it is possible that factors such as salinity, pH, alkalinity, carbonate ion concentration or secondary calcification may have a significant influence on the Mg/Ca ratios (Meland et al. 2006). In addition, Rosenthal and Boyle (1993) pointed out that dissolution may significantly alter the Mg concentrations within foraminifera shells, hence may impact Mg/Ca ratio. Nevertheless, as long as only the most pristine tests are picked, the Mg content of many planktic foraminifers can be used as a tool to trace the paleotemperature (Nürnberg 1995; Aagaard-Sørensen et al. 2013), keeping in mind that Mg incorporation might be species specific (Kristjásdóttir et al. 2007). Kozdon et al. (2009) proposed a reliable equation (for temperatures above ~3 °C) based on *N. pachyderma* (sin.) from the Nordic Seas that is being used today for the reconstruction of past SST, e.g., in the Fram Strait (Spielhagen et al. 2011; Aagaard-Sørensen et al. 2013):

$$\mathrm{Mg/Ca}\ \left(\mathrm{mmol}^{\mathrm{mol}^{-1}}\right) = 0.13(\pm 0.037) * \mathrm{T} + 0.35(\pm 0.17)$$

Recently, scientists began calibrating the Mg/Ca ratio for the reconstruction of past bottom water temperatures (BWT) based on benthic foraminifera (Skirbekk et al. 2012) and ostracods (Farmer et al. 2012). The latter appear to be better proxies because the Mg/Ca ratios in their tests are positively and linearly correlated to the BWT in the North Atlantic Ocean, the Nordic Seas and the Arctic Ocean.

2.4 Fragmentation/Dissolution

Planktic foraminifera shells can be exposed to carbonate dissolution that is associated with ocean circulation and climate (e.g., Archer and Maier-Reimer 1994; Archer 1996). Dissolution may occur in the water column as well as at the sediment-water interface and in the sediments (e.g., Lohmann 1995). Hence, the shell weight of the planktic foraminifera specimens can be used as a measure of the preservation of calcium carbonate (e.g., Broecker and Clark 2001; Barker and Elderfield 2002; Moy et al. 2009). *N. pachyderma* (sin.) is the dominant species in cold waters of the Arctic region. It displays a linear increase in abundance with decreasing temperature and reaches 100 % of the planktonic fauna in near-freezing waters (Kohfeld et al. 1996). Because of its abundance and better resistance, it is broadly used in dissolution determinations (e.g., Zamelczyk et al. 2012; Berben et al. 2014).

Only unbroken shells with no sediment fill and without signs of corrosion and indications of secondary calcite crusts can be used for dissolution analysis. Furthermore, the fragmentation of foraminiferal tests also reflects the degree of dissolution (Conan et al. 2002) and is inversely correlated with the shell weight (Berben et al. 2014). The fragmentation is calculated using the equation of Pufhl and Shackelton (2004):

$$Fragmentation\ (\%) = \frac{no.\,fragments\ g^{-1}}{\frac{no.\,fragments}{3} + no.\ test\ g^{-1}} * 100$$

The total number of fragments per sample is divided by three because it is assumed that each shell breaks into more than one fragment (Pufhl and Shackleton 2004; Berben et al. 2014). The increase in shell weight and concomitant decrease in fragmentation is interpreted as a sharp improvement in preservation conditions (Zamelczyk et al. 2012; Berben et al. 2014).

Few mechanisms cause the dissolution of calcareous material in the Arctic region. Brine rejection at the marginal ice zone (MIZ) produces CO_2-rich and corrosive bottom water masses (e.g., Steinsund and Hald 1994). In addition, an increased accumulation of unutilised organic material is observed at the MIZ (Huber et al. 2000), which causes a decrease in the pH of the bottom waters (Scott et al. 2008). However, $CaCO_3$ is preserved better under permanent sea ice cover where the production of organic matter is low (Scott et al. 2008).

The intensified dissolution of planktic foraminifera in the Early Holocene that was recorded in the sediment core from the central Fram Strait was attributed to the proximity of the MIZ (Zamelczyk et al. 2012). In general, poor preservation conditions can be linked to the increased influence of Arctic water (Zamelczyk et al. 2012), while ameliorated preservation conditions are linked to lower organic matter productivity and a greater rain of $CaCO_3$, which are both related to the Atlantic surface water (Huber et al. 2000; Henrich et al. 2002). Furthermore, the solubility of $CaCO_3$ increases with decreasing temperature and increasing salinity, CO_2 concentration (Edmond and Gieskes 1970), and depth due to pressure (Archer and Maier-Reimer 1994).

2.5 Magnetic Susceptibility (MS)

Magnetic susceptibility (MS) is the degree of magnetisation of a material in response to an applied magnetic field. MS is controlled by the type and amount of magnetic minerals in a rock. It is sometimes dominantly controlled by paramagnetic minerals (mafic silicates, such as olivine, pyroxenes, amphiboles, micas, tourmaline, and garnets), often by ferromagnetic minerals (iron oxides or sulphides, such as magnetite and/or pyrrhotite, respectively) and much less frequently by diamagnetic minerals (calcite and quartz).

The magnetic susceptibility of ocean sediments in the Arctic is dependent on changes in both oceanography (e.g., Rasmussen et al. 1996, 1998; Kissel et al. 1997; Moros et al. 2002) and glacial activity (e.g., Hillenbrand et al. 2009). On glaciated margins, mass transportation (e.g., Robinson et al. 2000; Kuijpers et al.

2001; Rasmussen et al. 2007) and meltwater plumes (e.g., Lekens et al. 2005; Rasmussen et al. 2007; Łącka et al. 2014) significantly influence the MS. Therefore, MS has been used for lithostratigraphic correlations (e.g., Antoniades et al. 2011; Ojala et al. 2014), reconstructions of sedimentation patterns (e.g., Helmke et al. 2005; Lekens et al. 2006; Jessen et al. 2010), reconstructions of ice sheet dynamics (Antoniades et al. 2011), or recently to trace palaeo-methane emissions off western Svalbard (Johnson et al. 2014).

2.6 X-Ray Fluorescence (XRF)

X-ray fluorescence core scanning (XRF) was first developed at the Netherlands Institute for Sea Research (NIOZ) in 1988 (Jansen et al. 1998). It is a nondestructive, nearly continuous, and relatively fast analytical method of the down-core variability in the elemental composition of sediments from aluminium to uranium (Richter et al. 2006). It is based on the principle that X-rays collide with matter, which leads to the generation of secondary radiation, i.e., fluorescence. This makes it possible to determine the qualitative and quantitative element compositions of solids, liquids and powders.

XRF can provide high-resolution palaeoenvironmental information in a variety of sedimentary settings. It has been applied to broad palaeoceanographic reconstructions, such as initial correlations between cores, preliminary stratigraphic interpretations of sedimentary sequences (e.g., Norris and Röhl 1999; Pälike et al. 2001), investigations of terrigenous input patterns (e.g., Croudace et al. 2006) and the provenance of terrigenous material (e.g., Haug et al. 2001; Lamy et al. 2004), tracing early diagenetic processes (Funk et al. 2004), spectral analyses of Milankovitch orbital cycles and sedimentation rate analysis (Peterson et al. 2000; Pälike et al. 2001, 2008). In the Arctic Ocean, it has been used to reconstruct the Palaeogene (Spofforth et al. 2008; Hanslik et al. 2013) and Late Quaternary (Polyak et al. 2009) stratigraphy and the glacial-interglacial variability of manganese content (Löwemark et al. 2008).

The interpretation of XRF results should be based on elemental ratios instead of single elements (e.g., Weltje and Tjallingii 2008). The most commonly used ratios for palaeoclimate reconstructions in the Arctic region are iron/calcium (Fe/Ca), titanium/calcium (Ti/Ca), titanium/aluminium (Ti/Al), iron/potassium (Fe/K) and aluminium/silicon (Al/Si) (Govin et al. 2012; see Table 1). Fe and Ti vary with the terrigenous fraction of the sediment because they are related to the siliciclastic components of the sediment (e.g., Arz et al. 1998; Jansen et al. 1998). In contrast, the coarse sediment fractions are enriched in Ti (e.g., Schütz and Rahn 1982; Shiller 1982), while Al and Si are mostly associated with fine-particle clay minerals (Biscaye 1965). Hence, the Si/Fe ratio is regarded as a proxy for the supply of sediments from glacial meltwater (Erbs-Hansen et al. 2013). Ca often represents the carbonate content of the sediment, the biological productivity of the surface water, the rate of dissolution during its journey through the water column and on the

Table 1 Examples of XRF parameters with their interpretations (after Croudace et al. 2006; Spofforth et al. 2008)

Measured property	Interpretation
Ca/Fe	• Indicative of the biogenic carbonate: detrital clay ratio • May be strongly correlated with sedimentary units • Can distinguish foraminifer- or shell-rich layers
Sr/Ca	• Enhanced Sr may indicate the presence of high-Sr aragonite, which requires a shallow-water source
Zr/Rb, Ti/Rb	• Useful as sediment-source/provenance indicators • Zr and Ti may be enhanced in turbidite bases
Si	• An important terrigenous or productivity indicator • May be strongly correlated with sedimentary units
Ba/Ti	• An important productivity indicator
Br/Cl, S/Cl	• High ratios may indicate organic-rich layers because Br and S are high in organic-rich sediments
Al, Ti, K, Fe	• Terrigenous indicators
Sr, Ca, Zr, Ba	• Marine productivity indicators
Ti/Al, K/Al	• Measures of sediment provenance • Measures of changes in energy (e.g., via water depth changes or from an accumulation of heavier particles)
Mn/Al	• Oxygenation level of the water column

seafloor, and the dilution by the non-carbonate fraction (Peterson et al. 2000; Bozzano et al. 2002; Jansen et al. 1998; Richter et al. 2006; Łącka et al. 2014). During interglacial conditions, this concentration is mainly controlled by the productivity of carbonate secreting organisms, such as foraminifera and coccolithophores (Hebbeln et al. 1998; Matthiessen et al. 2001). Ca/K and Ca/Ti ratios may be used as a preliminary down-core overview of sedimentary variability and for stratigraphic correlations between unprocessed marine cores (Hennekam and de Lange 2012). MS and Fe records exhibit similar patterns (i.e., high MS corresponds to high Fe intensity) because both signals reflect the concentration of iron (oxide)-bearing minerals (Westerhold et al. 2008; Łącka et al. 2014).

3 Summary

In recent years, numerous new proxies have been developed for the reconstruction of past environmental conditions in the Arctic region. Some (e.g., IP_{25} and PIP_{25}) have been created specifically for this region, while others have been successfully adapted to the polar regions after additional calibrations. For example, U_{37}^{K} was used in polar waters in the 1990s but gave false results with anomalously high temperatures during the Last Glacial Maximum (Rosell-Melé and Comes 1999), while it is currently used successfully in the Arctic. However, several established proxies require further calibration, such as TEX_{86} (Ho et al. 2014). Therefore, there is a large demand for research in the field of organic geochemistry and genetics.

The future of palaeoceanography depends on the development of new analytical techniques and the improvement of existing proxies. Classical methods will be replaced with new, highly efficient molecular and geochemical analyses that are independent of diagenetic processes and that provide continuous high-resolution records.

Acknowledgments This review paper was developed within the framework of grant no. 2012/05/N/ST10/03696 and 2011/01/N/ST10/06533 funded by the National Science Centre in Kraków (Poland).

References

Aagaard-Sørensen SA, Husum K, Hald M, Marchitto T, Godtliebsen F (2013) Sub Sea surface temperatures in the polar north Atlantic during the Holocene: Planktic foraminiferal Mg/Ca temperature reconstructions. Holocene 24(1):93–103

ACIA: Arctic Climate Impact Assesment (2005) Cambridge University Press, Cambridge

Alonso-Garcia M, Andrews JT, Belt ST, Cabedo-Sanz P, Darby D, Jaeger J (2013) A comparison between multiproxy and historical data (AD 1990–1840) of drift ice conditions on the east Greenland Shelf (~66°N). Holocene 23(12):1672–1683

Anderson-Carpenter LL, McLahlan JS, Jackson ST, Kuch M, Lumibao CY, Poinar HN (2011) Ancient DNA from lake sediments: bridging the gap between paleoecology and genetics. BMC Evol Biol 11:30. doi:10.1186/1471-2148-11-30

Andresen CS, Sicre M-A, Straneo F, Sutherland DA, Schmith T, Ribergaard MH, Kuijpers A, Lloyd JM (2013) A 100-year long record of alkenone-derived SST changes by southeast Greenland. Cont Shelf Res 71:45–51

Andrews JT, Belt ST, Olafsdottir S, Massé G, Vare L (2009) Sea ice and marine climate variability for NW Iceland/Denmark Strait over the last 2000 cal. yr BP. Holocene 19:775–784

Antoniades D, Francus P, Pienitz R, St-Onge G, Warwick FV (2011) Holocene dynamics of the Arctic's largest ice shelf. PNAS 108(47):18899–18904

Archer DE (1996) An atlas of the distribution of calcium carbonate in sediments of the deep sea. Global Biogeochem Cycle 10(1):159–174

Archer D, Maier-Reimer E (1994) Effect of deep-sea sedimentary calcite preservation on atmospheric CO_2 concentration. Nature 367:260–263

Arz HW, Pätzold J, Wefer G (1998) Correlated millennial-scale changes in surface hydrography and terrigenous sediment yield inferred from last-glacial marine deposits off northeastern Brazil. Quat Res 50(2):157–166

Axford Y, Andresen CS, Andrews JT, Belt ST, Geirsdóttir Á, Massé G, Miller GH, Ólafsdóttir S, Vare LL (2011) Do paleoclimate proxies agree? A test comparing 19 late Holocene climate and sea-ice reconstructions from Icelandic marine and lake sediments. J Quat Sci 26:645–656

Barker S, Elderfield H (2002) Foraminiferal calcification response to glacial-interglacial changes in atmospheric CO_2. Science 297(5582):833–836

Barker S, Cacho I, Benway H, Tachikawa K (2005) Planktonic foraminiferal Mg/Ca as a proxy for past oceanic temperatures: a methodological overview and data compilation for the Last Glacial Maximum. Quat Sci Rev 24(7–9):821–834

Barnett TP, Hasselmann K, Chelliah M, Delworth T, Hegerl G, Jones P, Rasmusson E, Roeckner E, Ropelewski C, Santer B, Tett S (1999) Detection and attribution of recent climate change: a status report. Bull Am Meteorol Soc 80:2631–2659

Belt ST, Müller J (2013) The Arctic Sea ice biomarker IP_{25}: a review of current understanding, recommendations for future research and applications in Palaeo Sea ice reconstructions. Quat Sci Rev 79:9–25

Belt ST, Allard WG, Massé G, Robert JM, Rowland SJ (2000) Highly branched isoprenoids (HBIs): identification of the most common and abundant sedimentary isomers. Geochim Cosmochim Acta 64(22):3839–3851

Belt ST, Massé G, Rowland SJ, Poulin M, Michel C, LeBlanc B (2007) A novel chemical fossil of Palaeo Sea ice: IP_{25}. Org Geochem 38:16–27

Belt ST, Massé G, Vare LL, Rowland SJ, Poulin M, Sicre M-A, Sampei M, Fortier L (2008) Distinctive ^{13}C isotopic signature distinguishes a novel sea ice biomarker in Arctic sediments and sediment traps. Mar Chem 112:158–167

Belt ST, Vare LL, Massé G, Manners H, Price J, MacLachlan S, Andrews JT, Schmidt S (2010) Striking similarities in temporal changes to seasonal sea ice conditions across the central Canadian Arctic Archipelago during the last 7,000 years. Quat Sci Rev 29(25–26):3489–3504

Bendle J, Rosell-Melé A (2004) Distributions of U^{K}_{37} and $U^{K'}_{37}$ in the surface waters and sediments of the Nordic Seas: implications for paleoceanography. Geochem Geophys Geosyst 5:Q11013

Bendle JA, Rosell-Melé A, Ziveri P (2005) Variability of unusual distributions of alkenones in surface waters of the Nordic Seas. Palaeoceanography 20(2):PA2001

Berben SMP, Husum K, Cabedo-Sanz P, Belt ST (2014) Holocene sub-centennial evolution of Atlantic water inflow and sea ice distribution in the western Barents Sea. Clim Past 10:181–198

Bik HM, Sung W, De Ley P, Baldwin JG, Sharma J, Rocha-Olivares A, Thomas WK (2011) Metagenetic community analysis of microbial eukaryotes illuminates biogeographic patterns in deep-sea and shallow water sediments. Mol Ecol 21:1048–1059

Biscaye PE (1965) Mineralogy and sedimentation of recent deep-sea clay in the Atlantic Ocean and adjacent seas and oceans. Geol Soc Am Bull 76(7):803–832

Boere AC, Abbas B, Rijpstra WIC, Versteegh GJM, Volkman JK, Sinninghe Damsté JS, Coolen MJL (2009) Late-Holocene succession of dinoflagellates in an Antarctic fjord using a multi-proxy approach: paleoenvironmental genomics, lipid biomarkers and palynomorphs. Geobiol 7:265–281

Boere AC, Sinninghe Damsté JS, Rijpstra WIC, Volkman JK, Coolen MJL (2011a) Source-specific variability in post-depositional DNA preservation with potential implications for DNA based paleoecological records. Org Geochem 42:1216–1225

Boere AC, Rijpstra WIC, De Lange GJ, Malinverno E, Sinninghe Damsté JS, Coolen MJL (2011b) Exploring preserved fossil dinoflagellate and haptophyte DNA signatures to infer ecological and environmental changes during deposition of sapropel S1 in the eastern Mediterranean. Paleoceanography 26:PA2204

Bozzano G, Kuhlmann H, Alonso B (2002) Storminess control over African dust input to the Moroccan Atlantic margin (NW Africa) at the time of maxima boreal summer insolation: a record of the last 220 kyr. Palaeogeogr Palaeoclim Palaeoecol 183(1–2):155–168

Broecker WS, Clark E (2001) An evaluation of Lohmann's foraminifera weight index. Paleoceanography 16:531–534

Brochier-Armanet C, Bousssau B, Gribaldo S (2008) Mesophilic Crenarchaeota: proposal for a third archaeal phylum, the Thaumarchaeota. Nature Rev Microbiol 6:245–252

Brohan P, Kennedy JJ, Harris I, Tett SFB, Jones PD (2006) Uncertainty estimates in regional and global observed temperature changes: a new data set from 1850. J Geophys Res 111:D12106

Bradley RS (2014) Paleoclimatology: reconstructing climates of the quaternary (3rd edition). Elsevier/Academic Press, San Diego, p 675

Brassell SC (1993) Applications of biomarkers for delineating marine paleoclimate fluctuations during the Quaternary. In: Engel MH, Macko SA (eds) Organic geochemistry. Plenum, New York, pp 699–738

Brassell SC, Eglinton G, Marlowe IT, Pflaumann U, Sarnthein M (1986) Molecular stratigraphy: a new tool for climatic assessment. Nature 320:129–133

Brown TA, Belt ST, Mundy C, Philippe B, Massé G, Poulin M, Gosselin M (2011) Temporal and vertical variations of lipid biomarkers during a bottom ice diatom bloom in the Canadian

Beaufort Sea: further evidence for the use of the IP_{25} biomarker as a proxy for spring Arctic Sea ice. Polar Biol 34:1857–1868

Brown TA, Belt ST, Tatarek A, Mundy CJ (2014) Source identification of the Arctic Sea ice proxy IP_{25}. Nature Commun 5:4197

Cabedo-Sanz P, Belt ST, Knies, Husum K (2013) Identification of contrasting seasonal sea ice conditions during the Younger Dryas. Quat Sci Rev 79:74–86. doi:10.1016/j.quascirev.2012.10.028

Chave KE (1954) Aspects of the biogeochemistry of 1. Calcareous marine organisms. J Geol 62:266–283

Chen JL, Wilson CR, Tapley BD (2006) Satellite gravity measurements confirm accelerated melting of Greenland ice sheet. Science 313:1958–1960

Comiso JC, Parkinson CL, Gersten R, Stock L (2008) Accelerated decline in the Arctic Sea ice cover. Geophys Res Lett 35:L01703

Conan SMH, Ivanova EM, Brummer GJ (2002) Quantifying carbonate dissolution and calibration of foraminiferal dissolution indices in the Somali Basin. Mar Geol 182(3–4):325–349

Coolen MJL, Boere A, Abbas B, Baas M, Wakeham SG, Sinninghe Damste JS (2006) Ancient DNA derived from alkenone-biosynthesizing haptophytes and other algae in Holocene sediments from the Black Sea. Paleoceanography 21:1–17

Coolen MJL, Volkman JK, Abbas B, Muyzer G, Schouten S, Sinninghe Damste JS (2007) Identification of organic matter sources in sulfidic late Holocene Antarctic fjord sediments from fossil rDNA sequence analysis. Paleoceanography 22(2):PA2211. doi:10.1029/2006PA001309

Coolen MJL, Orsi WD, Balkema C, Quince C, Harris K, Sylva SP, Filipova-Marinovad M, Giosan L (2013) Evolution in the plankton paleome in the Black Sea from the Deglacial to Anthropocene. PNAS 110(21):8609–8614

Croudace IW, Rindby A, Rothwell RG (2006) ITRAX: description and evaluation of a new multi-function X-ray core scanner. In: Rothwell RG (ed) New techniques in sediment core analysis. Geological Society of London, London, pp 51–63

de la Torre JR, Walker JC, Ingalls A, Könneke M, Stahl D (2008) Cultivation of a thermophilic ammonia oxidizing archaeon synthesizing crenarchaeol. Environ Microbiol 10:810–818

De Rosa M, Esposito E, Gambacorta A, Nicolaus B, Bu'Lock J (1980) Effects of temperature on ether lipid composition of *Caldariella acidophila*. Phytochemistry 19:827–831

Dzvonik JP (1996) Alkenones as records of oceanic paleotemperatures: studies of Eocene and Oligocene sediments from the north, south and Equatorial Atlantic, M.S. dissertation, Indiana University, Bloomington

Edmond JM, Gieskes JMTM (1970) On the calculation of the degree of saturation of seawater with respect to calcium carbonate under in-situ conditions. Geochim Cosmochim Acta 34:1261–1291

Erbs-Hansen DR, Knudsen KL, Olsen J, Underbjerg JA, Sha L (2013) Paleoceanographical development off Sisimiut, west Greenland, during the mid- and late Holocene: a multiproxy study. Mar Micropaleontol 102:79–97

Fahl K, Stein R (2012) Modern seasonal variability and deglacial/Holocene change of central Arctic Ocean sea-ice cover: new insights from biomarker proxy records. Earth Planet Sci Lett 351–352:123–133

Farmer JR, Cronin TM, Dwyer GS (2012) Ostracode Mg/Ca paleothermometry in the north Atlantic and Arctic Oceans: evaluation of a carbonate ion effect. Paleoceanography 27(2):2212. doi:10.1029/2012PA002305

Farrimond P, Eglinton G, Brassell SC (1987) Alkenones in Cretaceous black shales, Blake-Bahama Basin, western north Atlantic. Org Geochem 10:897–903

Funk J, von Dobeneck T, Reitz A (2004) Integrated rock magnetic and geochemical quantification of redoxomorphic iron mineral diagenesis in Late Quaternary sediments from the Equatorial Atlantic. In: Wefer G, Mulitza S, Ratmeyer V (eds) The south Atlantic in the Late Quaternary: reconstruction of material budgets and current systems. Springer, Berlin, pp 239–262

Francis JA, Vavrus SJ (2012) Evidence linking Arctic amplification to extreme weather in mid-latitudes. Geophys Res Lett 39:L06801. doi:10.1029/2012GL051000

Gong C, Hollander DJ (1999) Evidence for differential degradation of alkenones under contrasting bottom water oxygen conditions: implications for paleotemperature reconstruction. Geochim Cosmochim Acta 63:405–411

Govin A, Holzwarth U, Heslop D, Ford KL, Zabel M, Mulitza S, Collins JA, Chiessi CM (2012) Distribution of major elements in Atlantic surface sediments (36°N–49°S): imprint of terrigenous input and continental weathering. Geochem Geophys Geosyst 13(1):Q01013

Hanslik D, Löwemark L, Jakobsson M (2013) Biogenic and detrital-rich intervals in central Arctic Ocean cores identified using x-ray fluorescence scanning. Polar Res 32:18386

Haug GH, Hughen KA, Sigman DM, Peterson LS, Röhl U (2001) Southward migration of the Intertropical Convergence Zone through the Holocene. Science 293(5533):1304–1308

Harada N, Shin KH, Murata A, Uchida M, Nakatani T (2003) Characteristics of alkenones synthesized by a bloom of *Emiliania huxleyi* in the Bering Sea. Geochim Cosmochim Acta 67:1507–1519

Hebbeln D, Henrich R, Baumann K-H (1998) Paleoceanography of the last interglacial/glacial cycle in the polar north Atlantic. Quat Sci Rev 17:125–153

Helmke JP, Bauch HA, Rohl U, Mazaud A (2005) Changes in sedimentation patterns of the Nordic Seas region across the Mid-Pleistocene. Mar Geol 215:107–122

Hennekam R, de Lange G (2012) X-ray fluorescence core scanning of wet marine sediments: methods to improve quality and reproducibility of high resolution paleoenvironmental records. Limnol Oceanogr Methods 10:991–1003

Henrich R, Baumann K-H, Huber R, Meggers H (2002) Carbonate preservation records of the past 3 Myr in the Norwegian-Greenland Sea and the northern north Atlantic: implications for the history of NADW production. Mar Geol 184(1–2):17–39

Herbert TD (2003) Alkenone paleotemperature determinations. In: Holland HD, Turekian KK (eds) The ocean and marine geochemistry. Treatise on geochemistry. Elsevier-Pergamon, Oxford, pp 365–390

Higuchi R, Bowman B, Freiberger M, Ryder OA, Wilson AC (1984) DNA sequences from the quagga, an extinct member of the horse family. Nature 312:282–284

Hillenbrand C-D, Kuhn G, Frederichs T (2009) Record of a Mid-Pleistocene depositional anomaly in west Antarctic continental margin sediments: an indicator for ice-sheet collapse? Quat Sci Rev 28:1147–1159

Ho SL, Mollenhauer G, Fietz S, Martínez-García A, Lamy F, Rueda G, Schipper K, Méheust M, Rosell-Melé A, Stein R, Tiedemann R (2014) Appraisal of TEX_{86} and TEX^{L}_{86} thermometries in subpolar and polar regions. Geochim Cosmochim Acta 131:213–226

Hoefs MJL, Versteegh GJM, Rijpstra WIC, de Leeuw JW, Sinninghe Damsté JS (1998) Postdepositional oxic degradation of alkenones: implications for the measurement of Palaeo Sea surface temperatures. Paleoceanography 13:42–49

Hollis CJ, Taylor KWR, Hadley L, Pancost RD, Huber M, Creech JB, Hines BR, Crouch EM, Morgans HE, Crampton JS, Gibbs S, Peasrson PN, Zachos JC (2012) Early Paleogene temperature history of the southwest Pacific Ocean: reconciling proxies and models. Earth Planet Sci Lett 349–350:53–66

Hopmans EC, Weijers JWH, Schefuß E, Herfort L, Sinninghe Damsté JS, Schouten S (2004) A novel proxy for terrestrial organic matter in sediments based on branched and isoprenoid tetraether lipids. Earth Planet Sci Lett 224:107–116

Houghton JT, Filho LGM, Callander BA, Harris N, Kattenberg A, Maskell K (1996) Climate change 1995: the science of climate change. Intergovernmental panel on climate change. Cambridge University Press, Cambridge, p 572

Huber R, Meggers H, Baumann KH, Henrich R (2000) Recent and Pleistocene carbonate dissolution in sediments of the Norwegian-Greenland Sea. Mar Geol 165:123–136

Huguet C, Kim J, Sinninghe Damsté JS, Schouten S (2006) Marine crenarcheaotal membrane lipids in decapods: implications for the TEX_{86} paleothermomether. Geochem Geophys Geosyst 7:Q11010

Huguet C, de Lange GJ, Gustafsson O, Middelburg JJ, Sinninghe Damsté JS, Schouten S (2008) Selective preservation of soil organic matter in oxidized marine sediments (Madeira Abyssal Plain). Geochim Cosmochim Acta 72:6061–6068

Huguet C, Kim J, de Lange G, Sinninghe Damsté JS, Schouten S (2009) Effects of long term oxic degradation on the TEX_{86} and BIT organic proxies. Org Geochem 40:1188–1194

IPCC (2007) Climate change 2007: impact, adaptation and vulnerability. Contribution of working group II to the fourth assessment report of the intergovernmental panel on climate change. Cambridge University Press, Cambridge

IPCC (2013) Climate change 2013: impact, adaptation and vulnerability. Contribution of working group I to the fifth assessment report of the intergovernmental panel on climate change. Cambridge University Press, Cambridge

Jansen JHF, Van Der Gaast SJ, Koster B, Vaars AJ (1998) CORTEX, a shipboard XRF-scanner for element analyses in split sediment cores. Mar Geol 151:143–153

Jessen SP, Rasmussen TL, Nielsen T, Solheim A (2010) A new Late Weichselian and Holocene marine chronology for the western Svalbard slope 30,000-0 cal years BP. Quat Sci Rev 29(9–10):130–1312

Johnson J, Phillips S, Panieri G, Sauer S, Knies J, Mienert J (2014) Tracking paleo-SMT positions using a magnetic susceptibility proxy approach from sediments on the Arctic Vestnesa Ridge, offshore western Svalbard. Geophys Res Abstr 16:EGU2014-13511 (EGU General Assembly)

Jones PD, Osborn TJ, Briffa KR (2001) The evolution of climate over the last millennium. Science 292:662–667

Karner M, Delong E, Karl D (2001) Archaeal dominance in the mesopelagic zone of the Pacific Ocean. Nature 409:507–510

Kim J-H, Schouten S, Hopmans EC, Donner B, Sinninghe Damsté JS (2008) Global sediment core-top calibration of the TEX_{86} paleothermometer in the ocean. Geochim Cosmochim Acta 72(4):1154–1173

Kim J-H, Huguet C, Zonneveld AF, Versttegh GJM, Roeder W, Sinninghe Damsté JS, Schouten S (2009) An experimental field study to test the stability of lipids used for the TEX_{86} and U^{K}_{37} palaeothermometers. Geochim Cosmochim Acta 73:2888–2898

Kim J, van der Meer J, Schouten S, Helmke P, Willmott V, Sangiorgi F, Koç N, Hopmans E, Sinninghe Damsté JS (2010) New indices and calibrations derived from the distribution of crenarchaeal isoprenoid tetraether lipids: implications for past sea surface temperature reconstructions. Geochim Cosmochim Acta 74:4639–4654

Kim J-H, Crosta X, Willmott V, Renssen H, Bonnin J, Helnke P, Schouten S, Sinninghe Damsté JS (2012) Holocene subsurface temperature variability in the eastern Antarctic continental margin. Geophys Res Lett 39:L06705

Kissel C, Laj C, Lehman B, Labyrie L, Bout-Roumazeilles V (1997) Changes in the strength of the Icelande Scotland overflow water in the last 200,000 years: evidence from magnetic anisotropy analysis of core SU90-33. Earth Planet Sci Lett 152:25–36

Kohfeld KE, Fairbanks RG, Smith SL, Walsh ID (1996) *Neogloboquadrina pachyderma* (sinistral coiling) as paleoceanographic tracers in polar oceans: evidence from northeast Water Polynya plankton tows, sediment traps, and surface sediments. Paleoceanography 11(6):679–699. doi:10.1029/96PA02617

Kozdon R, Eisenhauer A, Weinelt M, Meland MY, Nürnberg D (2009) Reassessing Mg/Ca temperature calibrations of *Neogloboquadrina pachyderma* (sinistral) using paired δ44/40Ca and Mg/Ca measurements. Geochem Geophys Geosyst 10:Q03005

Kristjásdóttir GB, Lea DW, Jennings AE, Pak DK, Belanger C (2007) New spatial Mg/Ca-temperature calibrations for three Arctic, benthic foraminifera and reconstruction of north Iceland shelf temperature for the past 4000 years. Geochem Geophys Geosyst 8(3):Q03P21. doi:10.1029/2006GC001425

Kuijpers A, Nielsen T, Akhmetzhanov A, de Haas H, Kenyon NH, van Weering TCE (2001) Late Quaternary slope instability on the Faeroe margin: mass flow features and timing of events. Geo-Mar Lett 20:149–159

Lamy F, Kaiser J, Ninnemann U, Hebbeln D, Arz HW, Stoner J (2004) Antarctic timing of surface water changes off Chile and Patagonian ice sheet response. Science 304:1959–1962

Leininger S, Urich T, Schloter M, Schwark L, Qi J, Nicol G, Prosser JI, Schuster SC, Schleper C (2006) Archaea predominate among ammonia-oxidizing prokaryotes in soils. Nature 442:806–809

Lejzerowicz F, Esling P, Majewski W, Szczucinski W, Decelle J, Obadia C, Arbizu PM, Pawlowski W (2013) Ancient DNA complements microfossil record in deep-sea subsurface sediments. Biol Lett 9:20130283. doi:10.1098/rsbl.2013.0283

Llirós M, Gich F, Plasencia A, Auguet J, Darchambeau F, Casamayor E, Descy J, Borrego C (2010) Vertical distribution of ammonia oxidizing crenarchaeota and methanogens in the epipelagic waters of Lake Kivu (Rwanda Democratic Republic of the Congo). Appl Environ Microbiol 76:853–6863

Lekens WAH, Sejrup HP, Haflidason H, Petersen GØ, Hjelstuen B, Knorr G (2005) Laminated sediments preceding Heinrich event 1 in the northern north Sea and southern Norwegian Sea: origin, processes and regional linkage. Mar Geol 216:27–50

Lekens WAH, Sejrup HP, Haflidason H, Knies J, Richter T (2006) Meltwater and ice rafting in the southern Norwegian Sea between 20 and 40 calendar kyr B.P.: implications for Fennoscandian Heinrich events. Paleoceanography 21(3):PA3013. doi:10.1029/2005PA001228

Lohmann GP (1995) A model for variation in the chemistry of planktonic foraminifera due to secondary calcification and selective dissolution. Paleoceanography 10(3):445–457. doi:10.1029/95PA00059

Löwemark L, Jakobsson M, Mörth M, Backman J (2008) Arctic Ocean Mn contents and sediment color cycles. Polar Res 27:105–113

Łącka M, Zajączkowski M, Forwick M, Szczuciński W (2014) Late Weichselian and Holocene paleoceanography of Storfjordrenna, southern Svalbard. Clim Past Discuss 10:3053–3095. doi:10.5194/cpd-10-3053-2014

Marlowe IT, Green JC, Neal AC, Brassell SC, Eglinton G, Course PA (1984) Long chain (n-C_{37-39}) alkenones in the Prymnesiophyceae. Distribution of alkenones and other lipids and their taxonomic significance. Br Phycol J 19:203–216

Massé G, Rowland SJ, Sicre M-A, Jacob J, Jansen E, Belt ST (2008) Abrupt climate changes for Iceland during the last millennium: evidence from high resolution sea ice reconstructions. Earth Planet Sci Lett 269(3–4):565–569

Matthiessen J, Baumann K-H, Schröder-Ritzrau A, Hass C, Andruleit H, Baumann A, Jensen S, Kohly A, Pflaumann U, Samtleben C, Schäfer P, Thiede J (2001) Distribution of calcarous, siliceous and organic-walled planktic microfossils in surface sediments of the Nordic Seas and their relation to surface-water masses. In: Schäfer P, Ritzrau W, Schlüter M, Thiede J (eds) The northern north Atlantic: a changing environment. Springer, Berlin, pp 105–127

Meland MY, Jansen E, Elderfield H, Dokken TM, Olsen A, Bellerby RGJ (2006) Mg/Ca ratios in the planktonic foraminifer *Neogloboquadrina pachyderma* (sinistral) in the northern north Atlantic/Nordic Seas. Geochem Geophys Geosyst 7:Q06P14. doi:10.1029/2005GC001078

Moros M, Kuijpers A, Snowball I, Lassen S, Bäckström D, Gingele F, McManus J (2002) Were glacial iceberg surges in the north Atlantic triggered by climatic warming? Mar Geol 192:393–417

Moy AD, Howard WR, Bray SG, Trull TW (2009) Reduced calcification in modern southern ocean planktonic foraminifera. Nature Geosci 2:276–280

Müller J, Massé G, Stein R, Belt ST (2009) Variability of sea-ice conditions in the Fram Strait over the past 30,000 years. Nature Geosci 2(11):772–776

Müller J, Wagner A, Fahl K, Stein R, Prange M, Lohmann G (2011) Towards quantitative sea ice reconstructions in the northern north Atlantic: a combined biomarker and numerical modelling approach. Earth Planet Sci Lett 306:137–148

Navarro-Rodriguez A, Belt ST, Knies J, Brown TA (2013) Mapping recent sea ice conditions in the Barents Sea using the proxy biomarker IP_{25}: implications for Palaeo Sea ice reconstructions. Quat Sci Rev 79:26–36

Norris RD, Röhl U (1999) Carbon cycling and chronology of climate warming during the Palaeocene/Eocene transition. Nature 401:775–778
Nürnberg D (1995) Magnesium in tests of *Neogloboquadrina pachyderma* sinistral from high northern and southern lattitudes. J Foram Res 25(4):350–368
Ojala AEK, Salonen V-P, Moskalik M, Kubischta F, Oinonen M (2014) Holocene sedimentary environment of a high-Arctic fjord in Nordaustlandet. Svalbard Pol Polar Res 35(1):73–98
Okada H, Honjo S (1973) The distribution of oceanic coccolithophorids in the Pacific. Deep-Sea Res 20:355–374
Okada H, McIntyre A (1979) Validation of *Florisphaera profunda* var. *elongata* (2). Int Nannoplankton Assoc (INA) Newsl 1:2
Ochsenreiter T, Selezi D, Quaiser A, Bonchosmolovskaya L, Schleper C (2003) Diversity and abundance of Crenarchaeota in terrestrial habitats studied by 16S RNA surveys and real time PCR. Environ Microbiol 5:787–797
Pääbo S (1985) Molecular cloning of ancient Egyptian mummy DNA. Nature 314:644–645
Pääbo S, Poinar H, Serre D, Jaenicke-Després, Hebler J, Rohland N, Kuch M, Krause J, Vigilant L, Hofreiter M (2004) Genetic analyses from ancient DNA. Ann Rev Gen 38:645–79
Pälike H, Shackleton NJ, Röhl U (2001) Astronomical forcing in Late Eocene marine sediments. Earth Planet Sci Lett 193:589–602
Pälike H, Spofforth DJA, O'Regan M, Gattacceca J (2008) Orbital scale variations and timescales from the Arctic Ocean. Paleoceanography 23:PA1S10. doi:10.1029/2007PA001490
Pawlowski J, Christen R, Lecroq B, Bachar D, Shahbazkia HR, Amaral-Zettler L, Guillou L (2011) Eukaryotic richness in the abyss: insights from pyrotag sequencing. Plos One 6(4) e18169. doi:10.1371/journal.pone.0018169
Pawłowska J, Lejzerowicz F, Esling P, Szczuciński W, Zajączkowski M, Pawlowski J (2014) Ancient DNA sheds new light on the Svalbard foraminiferal fossil record from the last millennium. Geobiol 12(4):277–288. doi:10.1111/gbi.12087
Pearson A, Huang Z, Ingalls A, Romanek C, Wiegel J, Freeman K, Smittenberg R, Zhang C (2004) Nonmarine crenarchaeol in Nevada hot springs. Appl Environ Microbiol 70:5229
Prahl EG, Wakeham SG (1987) Calibration of unsaturation patterns in long-chain ketone compositions for paleotemperature assessment. Nature 330:367–369
Prahl FG, Wolfe GV, Sparrow MA (2003) Physiological impacts on alkenone paleothermometry. Paleoceanography 18(2):1025. doi:10.1029/2002PA000803
Prahl FG, Popp BN, Karl DM, Sparrow MA (2005) Ecology and biogeochemistry of alkenone production at Station ALOHA. Deep-Sea Res I 52:699–719
Peterson LC, Haug GH, Hughen KA, Röhl U (2000) Rapid changes in the hydrologic cycle of the tropical Atlantic during the last Glacial. Science 290:1947–1950
Polyak L, Bischof J, Ortiz JD, Darby DA, Channell JET, Xuand C, Kaufmane DS, Løvlie R, Schneider DA, Eberl DD, Adler RE, Councili EA (2009) Late quaternary stratigraphy and sedimentation patterns in the western Arctic Ocean. Global Planet Change 68(1–2):5–17
Pufhl HA, Shackleton NJ (2004) Two proximal, high-resolution records of foraminiferal fragmentation and their implications for changes in dissolution. Deep-Sea Res I 51:809–832
Rasmussen TL, Thomsen E, van Weering TCE, Labeyrie L (1996) Rapid changes in surface and deep water conditions at the Faroe margin during the last 58,000 years. Paleoceanography 11:757–771
Rasmussen TL, Thomsen E, van Weering TCE (1998) Cyclic changes in sedimentation on the Faeroe Drift 53-9 kyr BP related to climate variations. In: Stoker M, Evans D, Cramp R (eds) Geological processes on continental margins: sedimentation, mass-wasting and stability, vol 129. Geological Society Special Publication, London, pp 55–267
Rasmussen TL, Thomsen E, Slubowska MA, Jessen S, Solheim A, Koç N (2007) Paleoceanographic evolution of the SW Svalbard margin (76 N) since 20,000 14C yr BP. Quat Res 67:100–114
Richter TO, Van der Gaast SJ, Koster B, Vaars A, Gieles R, De Stigter HC, de Haas H, van Weering TCE (2006) The Avaatech XRF Core Scanner: technical description and applications

to NE Atlantic sediments. In: Rothwell RG (eds) New techniques in sediment core analysis, vol 267. Geological Society Special Publication, London, pp 39–50
Robinson SG, Sahota JTS, Oldfield F (2000) Early diagenesis in north Atlantic abyssal plain sediments characterized by rock-magnetic and geochemical indices. Mar Geol 163:77–107
Robson JN, Rowland SJ (1986) Identification of novel widely distributed sedimentary acyclic sesterterpenoids. Nature 324:561–563
Rontani JF, Belt ST, Vaultier F, Brown TA, Massé G (2014) Autoxidative and photooxidative reactivity of highly branched isoprenoid (HBI) alkenes. Lipids 49(5):481–494. doi:10.1007/s11745-014-3891-x
Rosell-Melé A, Comes P (1999) Evidence for a warm Last Glacial Maximum in the Nordic Seas or an example of shortcomings in $U^{K}_{37'}$ and U^{K}_{37} to estimate low sea surface temperature? Paleoceanography 14:770–776
Rosell-Melé A, Weinelt M, Ko N, Jansen E, Sarnthein M (1998) Variability of the Arctic front during the last climatic cycle: application of a novel molecular proxy. Terra Nova 10:86–89
Rosenthal Y, Boyle EA (1993) Factors controlling the fluoride content of planktonic foraminifera: an evaluation of its paleoceanographic applicability. Geochim Cosmochim Acta 57:335–346
Rosell-Melé A, Eglinton G, Pflaumann U, Sarnthein M (1995) Atlantic core-top calibration of the U^{K}_{37} index as a sea-surface paleotemperature indictor. Geochim Cosmochim Acta 59:3099–3107
Rueda G, Fietz S, Rosell-Melé A (2013) Coupling of air and sea surface temperatures in the eastern Fram Strait during the last 2000 years. Holocene 23(5):692–698
Sachs JP, Pahnke K, Smittenberg R, Zhang Z (2013) Biomarker indicators of past climate. In: Elias SA (ed) The encyclopedia of quaternary science, vol 2. Elsevier, Amsterdam, pp 775–782
Schiermeier Q (2012) Ice loss shifts Arctic cycles. Nature 489:185–186
Schouten S, Hopmans EC, Schefuss E, Sinninghe Damsté JS (2002) Distributional variations in marine crenarchaeotal membrane lipids: a new tool for reconstructing ancient sea water temperatures? Earth Planet Sci Lett 204:265–274
Schouten S, Hopmans EC, Sinninghe Damsté JS (2004) The effect of maturity and depositional redox conditions on archaeal tetraether lipid paleothermometry. Org Geochem 35:567–571
Schouten S, Forster A, Panoto E, Sinninghe Damsté JS (2007) Towards calibration of the TEX_{86} palaeothermometer for tropical sea surface temperatures in ancient greenhouse worlds. Org Geochem 38:1537–1546
Schouten S, Rijpstra W, Durich-Kaiser E, Schubert C, Sinninghe Damsté J (2012) Distribution of glycerol dialkyl glycerol tetraether lipids in the water column of Lake Tanganyika. Org Geochem 53:34–37
Schütz L, Rahn KA (1982) Trace-element concentrations in erodible soils. Atmos Environ 16 (1):171–176
Schulz H, Schöner A, Emeis KC (2000) Long-chain alkenone patterns in the Baltic Sea-an ocean-freshwater transition. Geochim Cosmochim Acta 64:469–477
Scott DB, Schell T, Rochon A, Blasco S (2008) Modern benthic foraminifera in the surface sediments of the Beaufort Shelf, slope and Mackenzie Trough, Beaufort Sea, Canada: taxonomy and summary of surficial distributions. J Foram Res 38:228–250
Seki O, Kawamura K, Sakamoto T, Ikehara M, Nakatsuka T, Wakatsuchi M (2005) Decreased surface salinity in the Sea of Okhotsk during the last glacial period estimated from alkenones. Geophys Res Lett 32(8):L08710. doi:10.1029/2004GL022177
Serreze MC, Walsh JE, Chapin FSII, Osterkamp T, Dyurgerov M, Romanovsky V, Oechel WC, Morison J, Zhang T, Barry RG (2000) Observational evidence of recent change in the northern high-latitude environment. Clim Change 46:159–207
Shiller AM (1982) The geochemistry of particulate major elements in Santa Barbara Basin and observations on the calcium carbonate-carbon dioxide system in the ocean, PhD thesis, University of California, San Diego, p 197
Sikes EL, Volkman JK, Robertson LG, Pichon J-J (1997) Alkenones and alkenes in surface waters and sediments of the southern ocean: implications for paleotemperature estimation in polar regions. Geochim Cosmochim Acta 61(7):1495–1505

Sinninghe Damsté JS, Hopmans EC, Schouten S, van Duin ACT, Greenevasen JAJ (2002) Crenarchaeol: the characteristic core glycerol dibiphytanyl glycerol tetraether membrane lipid of cosmopolitan pelagic crenarchaeota. J Lipid Res 43:1641–1651

Skirbekk K, Marchitto TM, Hald M (2012) Preliminary results of new benthic foraminifera Mg/Ca temperature calibrations and reconstruction of bottom water temperatures and salinity from a transect along the northern branch of the north Atlantic Current. AGU Fall Meeting, San Francisco

Sluijs A, Schouten S, Pagani M, Woltering M, Brinkhuis H, Sinninghe Damsté JS, Dickens GR, Huber M, Reichart G-J, Stein R, Matthiessen J, Lourens Lj, Pedentchouk N, Backman J, Moran K (2006) Subtropical arctic ocean temperatures during the Palaeocene/Eocene thermal maximum. Nature 441:610–613

Spang A, Hatzenpichler R, Brochierarmanet C, Rattei T, Tischler P, Spieck E, Streit W, Stahl D, Wagner M, Schleper C (2010) Distinct gene set in two different lineages of ammonia-oxidizing archaea supports the phylum Thaumarchaeota. Trends Microbiol 18:331–340

Spielhagen R, Werner K, Aagaard-Sørensen S, Zamelczyk K, Kandiano E, Budéus G, Husum K, Marchitto TM, Hald M (2011) Enhanced modern heat transfer to the Arctic by warm Atlantic water. Science 331(6016):450–453

Spofforth DJA, Pälike H, Green, DRH (2008) Paleogene record of elemental concentrations in sediments from the Arctic Ocean obtained by XRF analyses. Paleoceanography 23(1):PA1S09

Stein R, Fahl K (2012) Biomarker proxy IP_{25} shows potential for studying entire Quaternary Arctic Sea-ice history. Org Geochem 55:98–102

Stein R, Fahl K, Matthiessen J (2014) Late Pliocene/Pleistocene changes in Arctic Sea-ice cover: biomarker and dinoflagellate records from Fram Strait/Yermak Plateau (ODP Sites 911 and 912). Geophys Res Abstr 16:EGU2014-6895 (2014 EGU General Assembly)

Steinsund PI, Hald M (1994) Recent carbonate dissolution in the Barents Sea: Paleoceanographic applications. Mar Geol 117:303–316

Stoynova V, Shanahan TM, Hughen KA, de Vernal A (2013) Insights into Circum-Arctic Sea ice variability from molecular geochemistry. Quat Sci Rev 79:63–73. doi:10.1016/j.quascirev.2012.10.006

Uda I, Sugai A, Itoh Y, Itoh T (2001) Variation in molecular species of polar lipids from *Thermoplasma acidophilum* depends on growth temperature. Lipids 36:103–105

Vare LL, Massé G, Gregory TR, Smart CW, Belt ST (2009) Sea ice variations in the central Canadian Arctic Archipelago during the Holocene. Quat Sci Rev 28(13–14):1354–1366

Vare LL, Massé G, Belt ST (2010) A biomarker-based reconstruction of sea ice conditions for the Barents Sea in recent centuries. Holocene 20:637–643

Volkman JK, Barrett SM, Blackburn SI, Sikes EL (1995) Alkenones in Gephyrocapsa oceanica: implications for studies of paleoclimate. Geochim Cosmochim Acta 59:513–520

Volkman JK, Eglinton G, Corner EDS, Forsberg TEV (1980) Long-chain alkenes and alkenones in the marine coccolithophorid *Emiliania huxleyi*. Phytochem 19:2619–2622

Walther G-R, Post E, Convey P, Menzel A, Parmesan C, Beebee TJC, Fromentin J-M, Hoegh-Guldberg O, Bairlein F (2002) Ecological responses to recent climate change. Nature 416:389–395

Wefer G, Berger W, Bijma J, Fischer G (1999) Clues to ocean history: a brief overview of proxies. In: Fischer G, Wefer G (eds) Use of proxies in paleocenography: examples from the south Atlantic, Springer, Berlin Heidelberg, pp 1–68

Weijers JWH, Schouten S, Hopmans EC, Greenevasen JAJ, David ORP, Coleman JM, Pancost RD, Sinninghe Damsté JS (2006) Membrane lipids of mesophilic anaerobic bacteria thriving in peats have typical archaeal traits. Environ Microbiol 8:648–657

Weijers JWH, Schouten S, van den Donker JC, Hopmans EC, Sinninghe Damsté JS (2007a) Environmental controls on bacterial tetraether membrane lipid distribution in soils. Geochim Cosmochim Acta 71:703–713

Weijers JWH, Schouten S, Sluijs A, Brinkhuis H, Sinninghe Damsté JS (2007b) Warm arctic continents during the Palaeocene-Eocene thermal maximum. Earth Planet Sci Lett 261:230–238

Weltje GJ, Tjallingii R (2008) Calibration of XRF core scanners for quantitative geochemical logging of sediment cores: theory and application. Earth Planet Sci Lett 274:423–438

Westerhold T, Röhl U, Raffi I, Fornaciari E, Monechi S, Reale V, Bowles J, Evans HF (2008) Astronomical calibration of the Paleocene time. Palaeogeogr Palaeoclimatol Palaeoecol 257 (4):377–403

Willerslev E, Hansen AJ, Binladen J, Brand TB, Gilbert MTP, Shapiro B, Bunce M, Wiuf C, Gilichinsky DA, Cooper A (2003) Diverse plant and animal genetic records from Holocene and Pleistocene sediments. Science 300:791–795

Willerslev E, Hansen AJ, Poinar HN (2004a) Isolation of nucleic acids and cultures from fossil ice and permafrost. Trends Ecol Evol 19:141–147. doi:10.1016/j.tree.2003.11.010

Willerslev E, Hansen AJ, Ronn R, Brand TB, Barnes I, Wiuf C, Gilichinsky D, Mitchell D, Cooper A (2004b) Long-term persistence of bacterial DNA. Curr Biol 14:R9-10

Willerslev E, Cappellini E, Boomsma W, Nielsen R, Brand TB, Hofreiter M, Bunce M, Dahl-Jensen D, Johnsen S, Steffensen JP, Bennike O, Schwenninger J-L, Nathan R, deHoog C-J, Alfimov V, Christl M, Beer J, Muscheler R, Barker J, Sharp M, Penkman KEH, Haile J, Taberlet P, Gilbert MTP, Casoli A, Campani E, Collins MJ (2007) Ancient biomolecules from deep ice cores reveal a forested southern Greenland. Science 317:111–113

Wuchter C, Schouten S, Coolen MJL, Sinninghe Damsté JS (2004) Temperature dependent variation in the distribution of tetraether membrane lipids of marine Crenarchaeota: implications for TEX_{86} paleothermometry. Paleoceanography 19(4):PA4028. doi:10.1029/2004PA001041

Xiao X, Fahl K, Stein R (2013) Biomarker distributions in surface sediments from the Kara and Laptev Seas (Arctic Ocean): indicators for organic-carbon sources and sea-ice coverage. Quat Sci Rev 79:40–52

Zamelczyk K, Rasmussen TL, Husum K, Haflidason H, de Vernal A, Ravna EJK, Hald M, Hillaire-Marcel C (2012) Paleoceanographic changes and calcium carbonate dissolution in the central Fram Strait during the last 20 ka. Quat Res 78(3):405–416

MIX
Papier aus verantwortungsvollen Quellen
Paper from responsible sources
FSC® C105338

If you have any concerns about our products,
you can contact us on
ProductSafety@springernature.com

In case Publisher is established outside the EU,
the EU authorized representative is:
Springer Nature Customer Service Center GmbH
Europaplatz 3, 69115 Heidelberg, Germany

Printed by Libri Plureos GmbH
in Hamburg, Germany